Solving Your Growth and Revenue Problem

Lonnie Ayers

978-1-7366760-7-3 (kindle)

978-1-7366760-8-0 (Paperback)

CONTENTS

Table of Figures

Tables

CHAPTER ONE

Introduction

MY DESTINATION

Like every business owner, I need to have a goal to work toward. One of the hardest goals to set has always been my revenue goal. Of course, as anyone can tell you, if you don't know where you're going, any road will do.

As a small SAP consultancy owner that also happens to be a Hubspot Inbound Marketing and Sales Agency, my expertise is in helping companies set and achieve growth and revenue goals, and then hit them. It is almost a given that every new Hubspot client I take on struggles to set and achieve target revenues.

It seems simple enough when someone ask the question, 'what's your goal' however, it is very difficult, if not impossible to answer, unless you have a framework with which to think about and answer the question.

Setting that goal, and ultimately, solving the Growth and Revenue problem, is what this book is about. Though I've worked with some of the world's largest companies, such as Saudi Aramco and the United Nations, I've also worked with some of the tiniest start-ups, and each has faced and continues to face the ever-present challenge of setting and achieving Revenue Targets that will support their Growth Goals.

In the first book in this series, "How to Dominate Any Market: Turbocharging Your Digital Marketing and Sales Results[1]" I laid out the nine core problems every business must solve. Why nine and more importantly, why those nine in particular? Because those were the ones I constantly encountered in every client I worked with, even though I was specifically hired to provide Inbound Marketing and Sales consulting services. On the other hand, they were generally the same problems I found within my SAP clients, no matter which industry they were in or what geographic location they happened to be in.

In the second book in the series, "Knowing What You Sell: The Key to Dominate Your Market[2]" I did a deep dive based on the tough problem of

---

[1] Available on Amazon: https://www.amazon.com/Dominate-Turbocharging-Digital-Marketing-Results-ebook/dp/B08B3BLJ2H/ref=sr_1_3?dchild=1&keywords=how+to+dominate+any+market&qid=1634547725&sr=8-3

[2] Available on Amazon: https://www.amazon.com/KNOWING-WHAT-YOU-SELL-

knowing what you sell. This seems like something that should be easy to solve, but sadly, every business I've worked with, whether established or start-up, struggles to answer this most basic of questions. Using 'real world' case studies from my own experience, I demonstrate how to unravel this critical problem. If you have not read it yet, then I recommend you start with it.

In this third book of the series, I want to provide you with the tools, techniques, methodology and strategy to solve your growth and revenue problem.

DOMINATE-ebook/dp/B0946YYDK2/ref=sr_1_1?dchild=1&keywords=Knowing+What+You+Sell%3A+The+Key+to+Dominate+Your+Market&qid=1634547907&sr=8-1

CHAPTER TWO

How Much do You Want to Make?

DO YOU HAVE REALISTIC EXPECTATIONS?

Have you ever been asked how much you make? Perhaps you've been asked something even more difficult to answer - how much do you want to make this year, and the next? If you're like most people, the answer is "as much as possible".

The question then becomes, how much is possible? If you run a business, of any size, this is a question you can and must answer. For without knowing the answer, you cannot set expectations, for yourself or for the people in your organization.

Perhaps you're just starting out, and last year, you made $100,000 Gross Revenue. It was a good start, after all, you were just starting out. Many 'start ups' don't make anything, while a few experience wild success. As a Hubspot Inbound Marketing and Sales Agency owner, one of the hardest questions my Hubspot Channel Account Manager ask me is what are my goals for the year? Though we talk about many different kinds of goals, the one they are most interested in is my revenue goal.

Why do they focus on that, almost to the exclusion of everything else we will have to work together on? It's actually a very straight forward question they're asking. They want to know how much revenue I am planning on bringing in so they know how much they can plan on delivering to Hubspot.

Over the many years we've been Hubspot partners, there have, of course, been many CAMs that come and go. Like all such transitions, there's usually a hand-off meeting where the old CAM introduces the new one to me and we begin the process, all over again, of understanding how this relationship will work.

During the most recent iteration of this process, the old CAM, a very nice lady, explained that I had a lifestyle business. That was a polite way of saying that I was not a big business of hers and consequently, not a high priority for her. Let me be upfront, my goal was and is, grow a big business, with an eventual sale prior to my retirement, which is on the horizon, as part of 'the plan'. I have a definite 'buy out' figure in mind, based on some very well researched numbers.

I believe those numbers are realistic, given our current size, pipeline size and average deal value. We've even had a number of potential suitors make inquiries. But day-to-day, I am focused on closing that next deal, delivering excellent service and running the business.

When it comes to your own business, do you have realistic expectations? Have you figured out your 'magic numbers'? Do you know how to set realistic expectations, using simple math? That's what this book will teach you. How to set realistic growth and revenue expectations using simple math, proven to work across all industries, if not always called the same thing. Let's get down to business.

Summary and 10 Lessons Learned

Knowing your numbers is critical to having and setting realistic growth and revenue expectations. Even if you have a 'lifestyle' business, you still need to understand how to set and achieve realistic goals:

- **Know Your Numbers:** By numbers, I mean what your AOV[3] and more importantly, your 'magic numbers' are, which we will be developing as you progress through this book.
- **Understand Your Sales Pipeline[4]:** This goes along with knowing your numbers; with digital technology, you should be able to model, measure and improve your Top-to-Bottom Sales and Marketing Funnel[5].
- **Set Annual Revenue Goals:** When you follow the process I am going to walk you through in this book, you will know this number.
- **Realistic Expectations:** In order to set realistic expectations, it's simply easier to do based on math than on whims or unfounded assumptions.
- **Lifestyle Business vs. Big Business:** Yes, it makes a difference whether you want to intentionally remain small or build something bigger. Keep in mind that many 'side gigs[6]', where a lot of lifestyle businesses get their start, have gone on to grow very large.
- **Base Goals on Real Delivery Capability:** You don't have infinite delivery capability. You need to understand how much you can actually deliver, based on easily modeled production capacity, whether you're a service-based business or a product-based business or a combination of the two.

---

[3] AOV or Average Order Value

[4] A sales pipeline is a simple visual representation of every prospect you have arranged or bucketed into the various stages you have defined as your sales pipeline

[5] I refer to it as a Sales and Marketing funnel, as they both work together, even if they sometimes don't appear to.

[6] Facebook was originally a small side gig.

- **Have It Ready to Sell:** Even if you never intend to sell your business, your best 'Current' situation is to make sure it is ready to be sold.

- **Run a Clean Set of Books:** In order to have it ready to sell, I recommend you have a CPA set up your books and prepare the 3 periodic financial statements[7] directly from your system, such as Quickbooks or SAP.

- **Know Your Cost to Serve:** This is closely related to running a clean set of books and having those financial statements. You need to know what it cost you to deliver a dollar of service or product.

- **Understand Your Break-Even Point:** This is probably one of the most critical elements of solving your Growth and Revenue Problem. You have to know and track what your break-even point is on whatever it is you sell.

---

[7] Income Statement, Cash Flow Statement and Balance Sheet

CHAPTER THREE

Revenue Target

TOP LINE vs.. BOTTOM LINE TARGET SETTING

The first step in setting realistic revenue targets is to do some basic math. Specifically, determine how much Gross Profit you want to make. This is not a Financial Accounting book, nor do I intend to teach you how you to read a set of financial statements in this book[8]. But I do need to make sure we're on the same sheet of music to walk you through the process of setting achievable growth and revenue and targets.

Annual Gross Revenue	Gross Profit (%)	Profit
$1,000,000	.01%	$10,000
	.05%	$50,000
	.10%	$100,000
	.15%	$150,000
	.20%	$200,000
	.25%	$250,000
	.35%	$350,000
	.45%	$450,000
	.50%	$500,000

Table 1. Profit as a Percent of Gross Revenue

[8] One good place to find a course on basic financial literacy is on LinkedIn, which offers training on a wide variety of subjects, of which accounting is just one of many.

Nothing Magic About $1,000,000

I've specifically picked one million as it is easy to do percentages with, and generally speaking, not that hard of a revenue target to hit for many small businesses. However, during any subsequent math I demonstrate throughout this book, you should feel free to add or subtract zero's as needed.

There are many ways to read this table, but the way I want you to learn to read it is as 'Sales Equivalent'. This term was first introduced to me in the book by Mack Hanan, "Consultative Selling, The Hanan Formula for High-Margin Sales at High Levels" and later in a more effective, highly simplified approach in the book by Larry Myler, "Indispensable by Monday" which, at one point, had a nice associated website with a workbench for computing the Sales Equivalent Business Case. It's now offline, but you can get it from my own website: Business Case Builder[9]

You Really Have Two Goals to Set

When it comes to setting a revenue goal, you can set either a Top Line Revenue Goal or Bottom Line Profitability Goal, or both. To keep the numbers manageable, meaning useful to a small business, I will walk you through how I do it for my small consultancy. But let me preface this by saying, it's about the same process I've used to help companies that generate billions in revenue, for example, SAP, where I worked for a very long time.

The difference between doing it for myself and my own business versus doing it as a Quota Carrying SAP Industry Principal[10], is that in the first instance, I 'own' the Top Line Revenue Goal while in the second, it was assigned to me by my organization. I will walk you through the target setting process for both, using real world numbers.

---

[9] Sales Equivalent: https://www.sapbwconsulting.com/justify-your-project-value-with-our-business-case-builder-tool

[10] SAP Industry Principal: This role 'owns' an industry within SAP and is responsible for ensuring a pre-defined revenue target is set and met on an annual basis for that industry. Quota means how much revenue I was responsible for in my role as an SAP Industry Principal.

Setting the Revenue Goal for My Own Business

In reality, prior to setting the goal for my own business, I first needed to decide whether to start my own business. To do that, of course, there were many motivating factors, but one factor above all was my consideration of whether or not I could at least match my previous salary, which was $300,000 but could rise to as high as $450,000 with bonus' or much more if the gods were smiling upon me. If you'll take a moment to look at Table 1. Profit as a Percent of Gross Revenue, and interpolate between the percentage profit lines, you'll see that in order to generate $300,000 in my pocket on $1,000,000 in annual revenue, I would need to achieve 30% Gross Profit, at a minimum. This is different from net profit, but for purposes of our example, let's just use Gross Profit Margin.

Using a Gross Profit of 30% and an annual revenue target of $1,000,000, I could now see the beginnings of my business plan. But first, I needed to understand how much I needed to generate each hour I worked. To do that, I use the U.S. governments 2080 hours[11] of available to work figure each year.

Revenue Target	Hours Available	Hourly Earnings
$1,000,000	2080	$481

Table 2. Beginning of a Business Plan

There's two ways to read this table. The first is as shown, in that you need you earn $481 per hour to generate $1,000,000. To arrive at that figure, you just take the (Revenue Target/Hours Available)*Hourly Earnings. The second way is to interpret it as how much you're worth to the company on an hourly basis.

---

[11] The reason for using this source for this specific number is because otherwise, invariably, someone will have another number they insist is the real number. When I bid on government contracts or evaluate proposals to the government, I use this number. You can use any other number you wise, and this approach still works.

A Few Complicating Details

Of course, there's always a few details to be aware of when it comes to these things. The first is that you won't actually be able to work 2080 hours a year. Whatever world you work in, if you work by the hour, you have to take into consideration your utilization rate (known as a "ute" rate). What's a utilization rate? It's how many billable hours you actually work. And there's always lots of hours that are unavoidably not billable.

For example, within the SAP Professional Services world, i.e., SAP Consulting, we generally used a figure of 66% of the time as our maximum ute rate. Where did all those hours go? There were generally a few well-defined time-buckets you could count on: vacation, training, travel, meetings, supporting marketing events and pre-sales meetings. Let's take a look at the numbers now, using our 66% utilization rate.

Revenue Target	Hours Available	Hourly Earnings
$1,000,000	1373	$728

Table 3. First Business Plan Refinement

Tougher to Achieve That Figure Now

Now many of you will naturally say, hey, me and my guys, we are going to work all the time. Only the big guys have those sorts of hour stealing activities going on. My operation will be lean and mean. It's a nice thought. It's just not realistic, especially if you want to work as a consultant whose job it is "to know" - to be the tip-of-the-spear.

I've seen this utilization rate issue everywhere I've worked, including, most importantly, with the USAF (United States Air Force) depot overhaul facility[12] located in San Antonio, Texas. We employed tens of thousands of people overhauling C-130 Hercules, B-52 StratoFortresses and C-5A Galaxy cargo aircraft and related engines and components. We got, at best, 6 productive hours a day out of our people, which we referred to as Output Per Paid Man-Day, or OPMD. We paid them for 8, but 6 was the most you

---

[12] United State Air Force Depot Overhaul is where the heaviest maintenance is performed on military aircraft. By heavy, I mean, up to and including removing and replacing wings or completely rebuilding an aircraft from scratch.

could count on. The reasons were varied, but the point was and is, you never get 100% out of a day, and this holds true for machines as well.

Here's the thing you should now start to realize about that $1,000,000 Top Line Revenue Goal. Unless you're a super specialist, you most likely aren't going to achieve it on your own[13]. It's going to take a team of people.

How Do You Do This as an Internal Person?

As I mentioned before, I had to compare what I was making working as an internal 'employee' for SAP versus striking out on my own. It was and is always a very tough call. But let's work with the numbers I've already introduced and establish the lay-of-the-land a bit. As an SAP Industry Principal, I was a quota carrying salesperson. I had not always been a salesperson, in fact, I had actually been a SAP Project Manager for nearly 20 years. As an SAP Project Manager, I, of course, managed the financials of my projects, including project profitability. As an SAP Industry Principal, I had to manage the profitability of my 'business', which consisted of not just projects, but of software license sales.

The difference between this type of role and running my own business was mostly down to the number of resources available to me. I had a team of Account Executives who worked on my opportunities, and I had access to whatever other resources I might decide I needed, such as competitive information, pipeline information and other inputs.

Now that you know a bit about the SAP Industry Principal role, let's take a look at my numbers so we can eventually get to the decision-making process I had to go through when deciding to strike out on my own.

The SAP Industry Principal Numbers

At the very top level of my personal metrics was my target annual revenue. This is how much I had to sell in order to hit my OTE or On-Target-Earnings. At this level of sales, I could count on receiving 100% of my annual bonus number. In percentage terms, it came out to about 21%, which seems like a pittance but was not bad.

---

[13] There are, exceptions, of course.

You should know that in a typical year, I typically could achieve about 500% of annual revenue. Though there was no cap on the bonus, there were practical limits to my ability to pursue much beyond this number. Below is a table with semi-realistic numbers from that job.

Annual Revenue Target	Average Order Value (AOV)	Deals Required to Hit Target
$4,400,000	$100,000	44

Table 4. AOV vs. Deals Required to Hit Target

Let's see what the table looks like with a target of 500%.

Annual Revenue Target @500%	Average Order Value	Deals Required to Hit Target
$22,000,000	$100,000	220

Table 5. AOV vs. Deals Required to Hit Target @500%

If you do the math, you'll see that the job paid very well, as long as you actually hit those numbers and, here's the kicker, as long as the company honored its side of the deal. For those with a high-ticket sales background, you'll notice this is a linear model, with no real kicker to help you get there. The reader should also keep in mind that the T&Cs of the Sales Compensation program, which ran to 80 pages, were subject to annual changes, and you really couldn't negotiate. It was their way, or the highway. At some point, almost everyone decides on the highway. That is by design and works fine.

What's of much more importance is the number of deals you have to close to hit this target. Remember, I stated earlier that I had Account Executives who worked for me. In reality, I had parts of their time, as they supported every SAP Industry Principal. That's one of the challenges you will have when working in this type of model - lack of control.

Then there's the AOV or Average-Order-Value. In this case, the number was actually shrinking, year in and year out. That's why to make this happen, I had to employ a deeply stratified, highly segmented, partner strategy. You see, the reason I was called an Industry Principal, was that I have actual work experience in these industries. You rarely find an account executive who has much actual work experience within industry. More usually, you find that they are 'fresh meat'; given 5 days of training, and then given a bunch of accounts, and all the support the company could muster to help them close their deals.

Even being fresh, they still were very, very well paid and could earn as much as me, if they hit their numbers. What were those numbers? Please keep in mind these numbers change over time and companies have different sales compensation models.

Annual Target Revenue	Base Salary	Bonus (OTE 100%)
$1,400,000	$140,000	$140,000

Table 6. Example Bonus @ OTE

There were and typically are, other components of the compensation package, such as medical insurance, life insurance and stock options. But for purposes of illustration, I am keeping the basic model[14] simple.

Once you decide how much a salesperson needs to generate to hit quota (which is a separate subject), you can then calculate how much that salesperson is worth to the company. One way to do that is take the annual target revenue and divide it by the number of available hours, which again, is considered to be 2080 by accountants and the U.S. government. In the example given, that works out to be $673 Per Hour. I've included a small table to illustrate this

[14] Much of the 'basic model' included other elements aside from revenue (though that was what was weighted heaviest). For example, customer satisfaction was a component that counted, as did whether or not the customer was referenceable.

Total Compensation	Annual Hours	Value to Company
$280,000	2080	$673 Per Hour

Table 7. Annual Target Revenue to Bonus Example

Another way of reading this is to say that your Salesperson needs to deliver $673 dollars of revenue to the company, every hour. You also have some key data points that allow you to model your organization.

AOV	Annual Target Revenue	Closed Deals Per Month Required
$100,000	$1,400,000	1.16

Table 8. AOV - Target Rev. vs.. Closed Deals PM

Sources of Revenue Break-Down

When it comes to setting a revenue target, every company is different. But, as you can readily find information on publicly traded companies, such as SAP, Oracle and Microsoft, let's see where a company like SAP gets its revenue:

- Software Sales
- Software Support
- Implementation Revenue
- Training

Though there are many other sources of income, those four buckets cover most of it. For illustration purposes, we'll assume that each of these buckets brings in exactly 25% of the annual revenue. The profitability within each of these sources of revenue varies drastically. Let's do some hypothetical business planning. Let's first shrink the size of SAP to just $1,000,000,000 in annual revenue (even though it is close to 100 times larger than that).

That means we expect to generate $250,000,000 in software license revenue. The following table shows the numbers on this.

Annual Revenue Target	Software License Sales %	Revenue From Software License Sales
$1,000,000,000	25%	$250,000,000

Table 9. Software as a Percent of Revenue Target

We can now estimate how many Account Executives we'll need to have on staff if we hope to hit these numbers. It's sort of just simple math.

AE Annual OTE	Company Annual Target Revenue	Number of AEs Needed
$1,400,000	$250,000,000	179

Table 10. Number of AEs Needed

We can also estimate the number of Industry Principals we're going to need. Please keep in mind that there is overlap here in terms of revenue. Both of these roles are tasked with achieving the same revenue target.

IP Annual OTE	Annual Target Revenue	Number of IPs Needed
$4,400,000	$250,000,000	56

Table 11. Number of IPs Needed

Shaping the Sales Pipeline

Of course, selling consist of losing a lot of deals. In my experience, we typically won about 25% of our deals, or 1 out of 4. It had started out as winning 1 out 3 years ago, but as competition increased, and clients became more discerning, it had slowly become 1 out of 4. That translates into a pipeline coverage number. What it means is that you have to have at least 4X pipeline coverage[15] at all times. That last part is critical to understanding why the job requires so much marketing effort.

Annual Target Revenue	Sales Pipeline Coverage Required @ 4X	New Opportunity Required Per Month
$250,000,000	$1,000,000,000	Varies by Consumption Rate

Table 12. Target Revenue vs. Ops Required PM

Pipeline Coverage gets Consumed

The table above tells you that in our scenario, where the company plans on generating $250,000,000 in annual revenue, there needs to be $1,000,000,000 in pipeline coverage. That's a fancy way of saying how much opportunity you have to put and keep in your pipeline.

The challenge is that pipeline gets consumed and new pipeline coverage must be obtained. This means there is a never-ending requirement for marketing to deliver more opportunity to the pipeline. Marketing within this context, actually requires effort from many team members. Most importantly, it means involvement by the Industry Principal, who plays a key role of bringing actual industry expertise to the conversation.

Now that you know that the AE needs to close about 1 deal a month to hit quota and wins only about 25% or 1 out of four deals, you have some of the elements you need to figure out your opportunity. What you need is a time element to finish the pipeline conversion equation. And here's where it gets tricky.

[15] Pipeline coverage is simply how much deal value you need in your pipeline at any given time to hit your revenue target.

You see, your AE is probably going to be able to work, at most, 2 deals a day. What does 2 deals actually mean? What does "work it" mean? In my experience, in the best of circumstances, it means working on 2 opportunities a day, maximum. Realistically, one per day is more likely. Working it can mean a bunch of different activities, but within the software world, it usually means either being there when Pre-Sales Demonstrations are being given, or when Pre-Sales Discovery Sessions are being conducted, or closing the deal, or any number of other types of engagements that happen within any given sales opportunity. There's also a huge amount of time spent preparing proposals, presenting them, incorporating changes, and finally, getting to a close.

Closing the Industry Principal Loop

When I began this chapter, I introduced you to the concept of Industry Principal, provided his numbers and then proceeded to walk you through a hypothetical one billion a year revenue software company. You have a basic understanding of what the numbers are for each member of the team, at least the sales side of the house. So now let's tie together the IP role and the AE role.

Industry Principal Target Revenue	Deals Required to Hit Target	Account Executives Required (FTE)
$4,400,400	44	4

Table 13. IPs vs. Deals vs. AEs Required

Closing in On One Million Compensation

When it came to evaluating my current reality to my future vision of making one million a year, you now have the same data I had. The difference between the Industry Principal role and the new role of business owner, was that one was working within an established organization with a huge number of resources (although almost always not enough), and my own small practice where I had very few resources, other than my expertise. Both options would yield nice revenue, but one would yield recurring revenue to me and the other would continue to make somebody else rich.

Both routes, however, got me to my goal.

Summary and 10 Lessons Learned

By focusing on a realistic profit target, you put some realistic boundaries on your goal setting. Setting growth and revenue targets, regardless of the type of business you are in, means understanding who has to deliver how much revenue and by when. The key lessons learned:

- **Be Able to Read Financial Statements:** Even if you can't read a P&L, BS or CF statement, you need to be able to read basic financial reports. Flying blind never works.

- **Convert Gross Profit to Dollar Figures:** It's always helpful to convert a percentage (or decimal) Gross Profit number into actual dollar figures. It helps to focus your efforts.

- **Understand 'Sales Equivalent's Meaning:** If you can convert your product's value into a Sales Equivalent number, you can build a business case without knowing too much else about your customer's business. But if you do know exactly how much of your customer's cost your product reduces or how much revenue it drives, you can quickly use this concept to improve your sales results.

- **Work Backwards from Total Gross Profit Target:** By working backwards from your gross profit target, you can develop detailed, individualized revenue targets for members of the sales team, even if it is just you.

- **Convert Sales Quota into Hourly Revenue Targets:** Every sales team member is responsible for generating a certain number of dollars per hour. To be realistic here, you have to take into consideration the utilization rate.

- **Sales Productivity Should Be a Primary Metric:** Like all roles, sales can only do so much in a given day. I've provided a model, but there are many others. Call Center based operations may well be able to make a hundred calls a day, while a complex sale, shepherded by an Account Executive, will only be able to make one or even fewer a day.

- **Marketing Bears Responsibility for Leads:** Though all leads are appreciated, you must ensure your marketing team knows how many leads they must deliver to sales in order for sales to deliver the revenue.

- **Set Realistic Deal Closure Rate Goals:** In many situations, for a lot of different reasons, a sales person will only be able to close a small number of deals in any given time period. About the only way you are going to improve that is either to improve your offer or simplify your sales closing process, preferably, both.
- **Know Deal Closures Required to Hit the Annual Target:** Knowing how many deals you must close during the year is important. But knowing how many each member of your team must close during any day, week, month, quarter or year is even more important. If all deals are closed during the last day of the quarter, you're being sandbagged.
- **Bigger Teams Required for Complex Sales:** The more complex your product or service is, the bigger the team required to bring those deals in. These team members each have a cost, and that cost needs to be tied back to revenue. The composition of these teams tends to change over time as selling strategies evolve. The people in these extended Virtual Account Teams (VAT)[16] tend to be more expensive, hard to find, and even harder to keep. But without them, you're unlikely to hit your growth and revenue goals.

---

[16] A VAT (Virtual Account Team) is a term used to denote all the different people who are involved in pursuing a deal. There are other names for this, such as extended team, etc. However, these sorts of scenarios exist in all kinds of different businesses. Ever have a house built? You've dealt with a type of VAT. Architect, General Contractor, Bank, Real Estate Agent, the list goes on. The main characteristic that distinguishes this type of team from other arrangements is that all are focused on that single deal and working together to close the deal, which sometimes takes a long time and cost a lot of money.

CHAPTER FOUR

Capacity Determination

In the previous chapter, I introduced you to some basic math behind setting Revenue Targets so you can hit your growth goals. I intentionally kept the focus narrow, excluding many of the requisite parts you'll need to hit those targets.

Again, sticking with the hypothetical one billion a year software company, who we've already determined will need 179 Account Executives and 56 Industry Principals to hit the revenue target of $250,000,000 a year, let's now size the rest of the team.

You're Going to Need Pre-Sales Demonstration People

Like just about all enterprise software, our software requires people to demonstrate it to customers before they can make a purchase decision. They work on the Pre-Sales side of the house. Other vendors often call theirs Sales Engineers but I have found when those names are used, the level of skill is usually lower. This is simply a function of the need of the role and does not reflect badly on the person filling that role, but rather, reflects the level of expertise needed.

For someone to demonstrate software successfully, there are three phases to the process:

- Needs analysis
- Demo preparation
- Demo delivery and Q&A

The first thing to notice is that this process looks very much like developing and delivering an educational session - that's because that's exactly what it is. You're educating a client with a highly customized class. Let's take a look at the math behind this.

The first thing to know is that the majority of demonstrations cover the three core SAP modules[17], Finance, Materials Management and Sales and Distribution. However, for very large deals, the ones that I personally almost always worked on, the demonstrations would often cover all 16 core modules, and industry specific solutions as well as partner add-ons. For

---

[17] I use SAP here simply because I have experience with it. I could just as easily have used Oracle, Microsoft, i2 Technologies or Hubspot, all of which have very similar characteristics when it comes to sizing a team.

these demonstrations, which almost always were for large 7 to 8 figure software deals, it is often necessary to have 2 or 3 people deliver each of the demonstrations. It can also take several weeks to deliver these demonstrations. This is caused by client availability as well as our ability to corral the demonstration team in the same location for the demonstration.

How Much Will This Demonstration Cost to Deliver?

Let's make the assumption you're only going to demonstrate the three core modules. That means 3 people for 3 hours to deliver the demo. However, these demo's are customized, and therefore, we will have to pay for the demo preparation. As a rule of thumb, it will take about 3 hours of preparation time for each hour of demonstration. Let's do the math and put it in a table.

Demonstration Module	Demo Preparation Time (Hours)	Demo Delivery Time (Hours)
Financial Management	3	1
Materials Management	3	1
Sales and Distribution	3	1
Totals	9	3

Table 14. Demonstration Delivery Times

Now we have to figure out how much we are going to spend on this demonstration. We'll make some educated guesses using a concept called Fully Loaded Cost (FLC).

Role	FLC	Total Cost
Financial	$150 PH	$600
Materials Management	$150 PH	$600
Sales and Distribution	$150 PH	$600
Industry Principal	$200 PH	$800
Account Executive	$180 PH	$720
Totals		$3320

Table 15. Demonstration Delivery Cost Estimate

That's probably about as cheap as you're going to get on one of the smallest projects you can sell. Now, just to put it in perspective, let's model out one of the largest ones I personally managed, which was for a large railroad and lasted 3 solid weeks. In this scenario, we actually delivered at least 4 hours of demonstrations every day, but on some days, we delivered 8 solid hours of hours. At the end of the day, we delivered a demonstration of the entire Railroad Solution Map. Let's see what that one costed out at.

Roll	FLC	Total Cost
16 Different Modules	$150 PH	$9,000
Industry Principal	$200 PH	$24,000

Account Executive	$180 PH	$21,600
Totals		$54,600

Table 16. Total Demonstration Delivery Cost Example

Seems like a large investment to make but given the eventual size of the deal, it was more than worth it. Of course, there was always the risk of not winning, but in this one, I had a high degree of confidence that we would win. Why, when, on average, we only won about 25% of the time? Because we had done our homework when it came to building a business case, which is critical when it comes to winning these high-dollar, high-value deals. In particular, I had used some of my own budget to fund a Value Engineering team to develop and deliver highly focused value messages for this client. The cost for this 'service' was around $25,000 in total.

You Must Size Your Operation to Support This Approach

In order to deliver those Value Engineering messages, we had to have a Value Engineering team ready to go. Essentially, we had an internal management consulting company that we could put to work doing the business case part of this sales process.

It should be becoming very apparent that I am talking about a very complex sales process. I am. But I can assure you that there are other, even more complex products and services out there that require even more investment to sell. For instance, weapons systems for the United States Government can take a decade or more of sustained 'selling' and thus the sinking of cost. There are also many products that require far less. But in ALL cases, there is a cost of sales you must be aware of, model and manage, or you won't win.

Recapping where we are at in sizing our operation, we first determined what our annual target revenue was. Then we figured out how many Account Executives, Industry Principals and Pre-Sales people we would need. Then we made some educated guesses based on my own 22 years of experience driving SAP sales, as to how many customers an AE could work effectively. As well, I gave you some numbers around what it would take to deliver a Pre-Sales Demonstration. I also dropped a number of hours per year number on you, 2080 and the concept of Fully Loaded Cost Rate. Sort of a giant word problem, sounds like.

Here's how it looks so far for the Account Executive Role

Annual Target Revenue (Quota)	AOV	Deals Per Year to Hit Quota	Win Rate	Deals Worked
$1,400,000	$100,000	14	25%	56

Table 17. Quota-to-Deals Worked Relationship

Here's how it looks for the demo team.

This assumes that all deals require a demonstration, and all will consist of the three core modules.

Demo's Delivered Per AE	Total Man-Hours Required.	Total AEs	Total Man-Hours Required.	Total Pre-Sales Staffing
56	672	179	120,288	58

Table 18. Demo's to Pre-Sales Staffing Req.

This is what it will take to work the sales deals to hit that one-billion-dollar mark. It does not include what it will cost to develop or market or implement the software. It also does not include the Value Engineering Team or in the case of SAP, what they call the Business Transformation Team[18]. As well, it does not include any management or other overhead.

[18] Business Transformation is technically a different team within this environment.

But roughly speaking, it means that for this particular kind of company, for every one billion in annual revenue, you're going to need 237 people just to work the deals. On top of that, you will need marketing, development, support, training and implementation capabilities, each of which has the same sorts of numbers you will need to work through to size them.

Of course, to size those operations, you need to get the data. That's what we do next.

However, I generally made use of both of their services on complex deals as they are addressing different parts of the problem, namely, value versus solution footprint.

Summary and 10 Lessons Learned

Sizing your operation is a critical task. It should be driven by your business plan, the complexity of your offering and your sales process. It is also limited by your ability to fund it. Most very small companies won't be able to build out this model I've laid out above, but can partner up to achieve the same goal:

- **Complex Solutions Require More Resources:** Very high-ticket complex solutions require more resources to market and sell. But even simple solutions can require more resources to sell than you're expecting.
- **Match Demonstration Team Roles to Customer Roles:** This one is a tough one. It is not often that a keyboard commando is safe to take lead on business discussions with a CEO. Conversely, the old guy who is the point guy for the 'business level' discussions is probably not the best person to run the keyboards, unless that's what they really do.
- **Recorded Demo's OK for Marketing Purposes:** It's generally ok to use recorded demo's for marketing purposes. Be warned - don't fall into the trap of just supplying the recorded demo to the client. I've yet to see a sale result when this approach was taken.
- **Sales Demo's Usually Need to Be Customized:** Most system demo's of complex systems i.e., SAP, Oracle, Microsoft, need to be customized to specific customer requirements. These are better delivered in a recorded fashion as 'live' demo's tend to go sideways.
- **Sales Demo's Can Be Delivered as Recordings.** By delivered, I mean by a person who is an expert with the module and uses the recording for the demonstration.
- **Estimate Demo Cost Prior to Proceeding:** You can spend more than you expect to make on a demonstration. Therefore, you need to use the rules of thumb I provided in the demo costing table. Your particular demonstration cost will be based on your own unique circumstances and capabilities. However, it will involve knowing

the FLC rate of each person involved in preparing it, as well as the 'past experience' you should have in preparing demonstrations.

- **Track Sales Costs to the Account Level.** If you want to know what you spent to close a deal, you have to create a deal, which should be linked to whatever you define as an account.
- **Different Systems Use Different Deal Models.** Each of the major CRM vendors has a slightly different view of the Account/Opportunity/Deal relationship.
- **Size Your Demo Team Properly:** There is nothing more wasteful than to have generated demand for which you have insufficient capacity to perform demonstrations.
- **Equip Your Demo Team with the Right Tools.** This can be especially difficult for smaller operations. You need working systems that are also suitable for pre-sales and sales. But your demonstration team also needs many other tools, such as video recording and editing tools. Your team, aside from being highly skilled with your product, also needs to be highly trained[19] in developing and delivering effective presentations and demonstrations.

[19] One of the smartest things I ever saw SAP do was to send a small team of Presales Demonstration experts to the Middle East to train up our presales team. They stayed for a long time and it made a huge difference in the outcome.

CHAPTER FIVE

Get the Data First

GETTING DATA YOU CAN TRUST

When it comes to data, today's businesses are overflowing with it. It is coming in from everywhere, in vast quantities. It should be easy to get whatever you might need, or so it would seem. The problem is that most of the time, the data is either missing, full or errors, or you don't have access to it.

That's why, at least on the digital marketing front, you need to have all of your analytics installed and tracking. That's typically going to consist of Google Analytics, Google Tag Manager, Google Adwords, and Google Search Console. If you're a customer of ours, you will be on Hubspot, which comes with the analytics tracking code you need to track basically everything someone does from when they first find you on the internet until the last day they are a customer of yours.

In addition to these two core sets of tracking codes, you're typically going to have code from Facebook, Twitter and LinkedIn, and probably a whole bunch of others[20]. Each requires some setup and each works slightly different. But if you want to know how each marketing channel is performing for you, then you will have to work through the sometimes frustrating setup process of all these.

What you're also going to find is you need to have analytics that ties all of your other systems together, i.e., ERP and other transactional systems. The bigger the business, the more systems you typically find; some banks I've worked with had over 1,000 systems. The US Government had at least 4,400 ERP class systems in the 1980s. I don't expect they've managed to retire them or greatly simplify their system environment.

What all of this system complexity means is that no matter how 'natively integrated' your systems theoretically are, there's going to be a lot of data that gets lost as it passes through your systems. There's usually no problem with how a system handles data within its own boundaries, but when data gets passed, whether via APIs, custom integrations, or more sophisticated approaches, like Enterprise Application Integrations or EAIs, when it comes to developing a mathematical model of your business, expect some bad data to be part of the mix.

---

[20] For a general idea of the potential complexity you may encounter, take a look at all of the standard Tags available within Google Tag Manager.

Maybe You'll Get Lucky

In the best of all worlds, you'll have a system like Hubspot Sales and Marketing as your core system. All of your leads will come in via Hubspot forms, be marketed to on a 100% Hubspot hosted system, and finally, turned into customers using Hubspot's CRM. In these cases, you can easily find the required conversion rate information you're going to need for some of the calculations we're going to work through a little later on.

In all likelihood, you're going to find yourself in a mixed environment. Some of your website will be on Hubspot, some will be on another CMS system like WordPress, and your CRM will be something else, perhaps Salesforce. All is not lost, you can integrate all of these systems together somewhat easily. But there will be glitches. For instance, you may run out of something called API calls between Hubspot and Salesforce. You will miss some of your lead flow and lose money as a result.

On the WordPress front, you'll almost certainly experience technical glitches on a near constant basis. Plug-ins will need to be updated, and when you do, you'll break something. Hackers will break into your system, and you'll need to work on security issues. I speak from experience[21].

You'll End Up Using Excel

Depending on your point of view, the fact you're going to end up using excel is either a good thing or a bad thing. But the bottom line is, you're going to end up using a spreadsheet to do this modeling, once you get the data you need and that you can trust. That's ok, this is simple math. And spreadsheets are fairly easy to use. There's plenty of tutorials out there for most any problem you'll come up with.

But it won't have the model you need or the data you require. That's what you need to develop and what I am going to be walking you through, as soon as you knock out one key activity, figuring out your Sales Pipeline stages.

---

[21] Though our WordPress Sites have the best security available and installed, I've seen entire sites taken down for over a month due to hackers. Always good to have a back-up. Don't be a soft target.

Summary and 10 Lessons Learned

Having data you can trust is key to your success. There's going to be way more data being generated than you know what to do with in most business scenarios. Getting data you can trust and that consistently yields actionable insights means:

- **Have a Business System Design.** Have an overall business systems design for both your business processes and IT Systems.
- **Have Trustworthy Analytics.** Having all of your web analytics tracking codes installed and tested end-to-end.
- **Data Dictionary.** Have a corporate wide data dictionary. Make sure everyone on every team has access to it and that all teams understand the meaning of each term and agrees with that meaning. Provide a mechanism that allows for updating definitions or creating new terms.
- **Know what a Lead and Customer is.** Everyone must agree on what terms such as lead, and customer actually means within the context of your business.
- **Situational Data Awareness.** Knowing what data you have and that which you do not have and cannot get.
- **Auditable Data Trail.** Having an audit trail you can follow for every piece of data.
- **Documented Integration Layer.** Having every integration point in your information chain fully documented and stress tested until the point of failure.
- **Tracking 3rd Party Changes to Analytics Code.** Knowing when changes have been made by outside entities that directly affect your business. For instance, Google makes fairly constant changes to its tracking code and this changes the meaning of the data you are seeing over time.
- **Understand Data Exchanges.** Knowing the limits of your data exchanges. For instance, if you have a system, such as Salesforce

integrated with a system such as SAP or Hubspot, what happens when you exceed your API[22] limits. What critical piece of data is affected, i.e., inventory balances may be completely out of whack yet sales keeps selling.

- **Security of Data.** Having your data secured, end-to-end, and keeping that security up-to-date.

[22] API - Application Programming Interface

CHAPTER SIX

Figure Out Pipeline Stages (If Missing)

KEEP IT SIMPLE

Your sales pipeline can be very simple or very complex, and though you may 'design' it, your customers ultimately define it by their actions. That's why when it comes to designing your Inbound Sales based Sales Pipeline, you need to focus on the steps your customer takes at each step, not the steps you take. We'll address those a little later on.

Sales Pipeline Design Step-by-Step

Before I walk you through my simple Sales Pipeline Design Process, I have to make clear a couple of assumptions:

- This is for a large dollar, multi product or service business
- It's mostly applicable to what is called a considered sale[23].

Why a multi-product or service? Because most companies of any size sell many products or provide many services. And the pipeline for any one of them may well need to have its own pipeline modeled within your CRM system. It's also to let you know that when a company has a very large number of products, then the pipeline information is often intentionally compressed for Sales Management purposes and this process would need to be modified.

What are the steps I would take in defining a multi-product sales pipeline? As has been my practice throughout this series of books, I will lean on both my SAP experience for the large enterprise approach and my experience with a multitude of Hubspot Inbound Marketing and Sales customers. For the large enterprises in the IT space, the process for figuring this out goes something like this:

[23] A considered sale is when the customer has to conduct research on a product or service.

Customer Activity	Your Activity	Step - Yes or No
Makes an Inquiry	Provide Information	Yes
Issues RFP	Respond	Yes
Answers Discovery Session Questions	Conduct Discovery Session	Possibly - This is where you possibly exit the deal
Attends Demo Sessions	Conducts Demos	Yes
Accepts Proposal	Provides Proposal	Yes
Sign-Contracts	Process Internally	Yes
Rejects Proposal	Process Internally	No

Table 19. Identifying Deal Pipeline Stages

This is what the basic process looks like no matter how complex the eventual deal becomes. Here's the kicker. This is the vendor point of view. The buyer often times has many more steps in their buying process than what is in the seller's sales process. For instance, there will almost always be a request to see reference customers[24]. There will usually be an opportunity to develop a business case. This process can take a while and often times, has inherent conflict points built into it.

[24] There are many different types of reference customers and reference customer visit objectives.

Other typical sales funnel steps include financing. There are really two different elements to the financing step:

- The **Buyers Credit Worthiness** and **Ability to Pay**
- The **Vendors Financial Ability** and **Ability to Deliver**

It is up to you whether you want to include these steps in your funnel. However, my experience has been that most CRMs do not include[25] much beyond a checkbox to say yes or no to either of these financial elements. For instance, for large dollar RFPs, there are often numerous financial instruments and commitments that must be met by the seller before a deal can be done. This can include performance bonds, usually for at least the full value of the project, and many other types of bonds or other mechanisms that require the vendor to tie up cash in escrow accounts. As these requirements are specified in the RFP, and are usually different from buyer to buyer, I have never found it feasible to add a generic finance step to the pipeline. Instead, I use tasks and document management systems to keep track of these elements.

Other elements that I typically exclude are insurance certification requirements. These can be numerous in nature, almost always different from buyer to buyer, and also very expensive and time consuming. It is also sometimes the case that you chose to bid but decline to meet all of their so called 'mandatory' insurance requirements. You may find that they're flexible with these and were just trying to shift risk to you without compensation.

In the above basic Sales Pipeline, I've defined a simple six-step Sales Funnel. You'll notice this is far from one-click as you might find on Amazon or a Clickfunnels[26] style approach. It also tells you that though you may have a highly digitized sales process, the larger the deal, the more likely your sales pipeline will have a degree of people involvement.

I will warn the reader that the pipeline I dealt with in my ERP days had 13 steps defined in it. Because we had such a high volume of leads coming

---

[25] There are exceptions to this. SAP CRM, which is really for large enterprises, has a concept called FSCM or Financial Supply Chain Management in it. Within that system, there is an extensive system to evaluate the credit worthiness of a buyer (though nothing of the seller as of the publishing date of this book).

[26] Clickfunnels is a integrated web based digital sales system.

in, one of our pipeline management goals was to qualify out prospects as soon as possible. Some of the steps or additional data points you might end up adding to your multiproduct or service pipeline could include:

- Segment (i.e., Target, Portfolio, Strategic, Volume)
- Country
- Territory
- Opportunity Status
- Sales Organization
- Account Executive
- Virtual Account Team (including externals)

One other critical element of designing your sales pipeline is time. You'll notice I only listed the possible steps in the pipeline above. I did not say how long a prospect should or would be in any stage of the pipeline. Your CRM will keep track of that for you, but it will just be counting the actual amount of time a deal is in each stage, not what you designed it be.

Designing the Single Offering Pipeline

Many of my Hubspot Inbound Marketing and Sales customers are smaller enterprises, usually from 3 to 20 million in annual revenues. This means they usually sell only a single product or service. That said, sometimes, even these small enterprises have a wide variety of products they actually could sell, but experience has shown that most tend to focus on a niche and try to fully exploit it.

That makes them an ideal fit for Inbound Sales and thus, they also need an Inbound Sales pipeline design. Though the sales pipeline design process is mostly the same, the outcome should be very simple if it is to work as expected. Most importantly, if you're using HubSpot's CRM and marketing platform, it should result in activities that can be automatically measured by the system, without the Sales Team doing any additional tracking work. As usual, I will call upon some past client use cases, with the names changed to protect the innocent.

In this case, I am going to walk you through a typical pipeline for a SAP partner. You might say, didn't you use SAP in the first example? Why yes, yes I did. But there's a world of difference between the pipeline for a global company that develops software used by almost every type of customer and a small partner that focuses mostly on delivering niche services with perhaps one or two products they sell and attach services to. The steps will look something like this:

Customer Activity	Your Activity	Step - Yes or No
Initial Contact	Collect Contact Information	Yes
Complete's Scoping Questionnaire	Walk Customer Through Questionnaire	Yes
Accepts Proposal	Submit Proposal	Yes
Closed - Won	Initiate Work	Yes
Closed - Lost	Record Reason	Yes

Table 20. Refining Your Deal Pipeline Design

As you can see, there's not a lot of difference, except for the focus. Whereas the large multi-product or multi-service pipeline has only a few more steps, the complexity behind each step is profoundly different from the single product or service. This is not to say there are not many steps that even the smallest customers may try to get you to perform before getting to a deal. Most typically, one of these 'smaller steps' would be the reference call.

The reason I point out the reference customer call is that this is where many deals go to die. Perhaps you don't have any reference customers. Or, as has increasingly been the case, many customers have been advised by their legal counsel to never provide a customer reference. There are a number of reasons for this, none of which seem very legitimate to me. One of the primary ones is that they may reveal competitive information. This may be a possibility, but I've set up dozens of customer reference visits and it has not been an issue.

When it comes to accepting your customer's request to provide a customer reference, there's a couple of 'Pro Tips' to keep in mind. First and foremost, make sure you know the reason the prospect wants to conduct a reference visit. Second, your prospect must provide the questions they are trying to get answers to from your customer. Lastly, never set one up 'in the blind', where you don't know who they will be talking to nor what questions they are asking.

Regardless of the number of steps in your Sales Pipeline, there will be a number you must know to figure out how much you will make and how much you can expect to make. We'll discuss that magic number next(s).

Summary and 10 Lessons Learned

Keeping it simple can be very complicated. Though they can look very similar, large enterprise, multi-product, multi-service pipelines are usually more complex than smaller enterprise pipelines. To keep your sales pipeline design effective, keep these lessons learned in mind:

- **Keep Your Pipeline Design Simple.** Know that buyers will almost universally try to make you sell the way they want to buy. If you find yourself tying yourself and your team into loops trying to meet a customer's convoluted buying process, it may be a good indicator they are not a good fit and you should qualify yourself out.
- **You Need Enough Capital to Compete.** Large enterprise sales usually tie up a lot of cash on the seller side, and this cash tie-up is driven by the buyer. Your system must allow you to easily keep track of how much, where and with whom you have this cash tied up.
- **Partner Effectively to Stay in the Game.** Smaller enterprises can be 'shut out' of these larger deals and usually are, unless they can partner effectively with the larger enterprise.
- **Run a Lean Pipeline.** Your sales pipeline design should be as simple as possible, meaning as few steps as possible should be in it.
- **Test Your Pipeline Automation.** Behind the scenes, each step in your sales pipeline induces additional IT complexity. Make sure you design and test this automation thoroughly. It should be invisible to the sales team.
- **Template Where You Can.** For large enterprise deals, where there will normally be complex but repetitive insurance and bonding requirements, it is advisable to get this to a 'repeatable template[27]' type process.

---

[27] Many industries, but especially the construction industry, will require massive sums of money to be 'locked away' in the form of surety bonds or other types of financial instruments. If you do business on a multinational basis, that means you need to have identified the local bank you will use and know all of their paperwork requirements for posting these various bonds. Most importantly, you need to know their fees for performing this service and as well, ensure the bank discloses exactly

- **Associate People and Companies Automatically.** You need a CRM system that allows you to tie contacts (people) to companies and opportunities - preferably automatically.
- **System Should Track Sales Activity Automatically.** Your CRM system should collect almost all data automatically, including phone calls made and received, emails sent and received, meetings scheduled and held and tasks scheduled and performed.
- **Know What Each Customer Has Bought.** Your pipeline management system should allow you to quickly see what has been sold to any given customer and what is about to be sold.
- **Integrate Your Sales Pipeline with Your ERP System.** Your CRM system and all associated pipelines need to be fully integrated with your ERP system or whatever you call your system for manufacturing and delivering your product or service.

---

how and to whom the fund will be released throughout the project.

CHAPTER SEVEN

Figure Out Conversion Rate Between Stages

JOURNEY TOWARDS THE MAGIC NUMBERS

Congratulations, you've done the hard work of defining your Sales Funnel! If you're just starting operations, you'll have to make some assumptions until you find your magic numbers. But, let's take a huge leap of faith and assume you already have sales coming in. As a matter of fact, you have a defined sales process, and prospects are starting to proceed through it with some regularity. That might look something like this after some months of operations:

Stage One - Inquiry	Stage Two – Demo	Stage Three – Buy
677	34	7
737	54	11
737	40	15
615	56	25

Table 21. Initial Sales Funnel Conversion Rates

Of course, your sales funnel will probably consist of several more stages. You will no doubt have multiple funnels as you have different products you sell.

What this table does not say is how many leads came into your marketing funnel before they made what you have decided qualifies as an inquiry. There are many ways a company can treat a lead before it drops into your sales funnel and officially becomes an inquiry.

The Funnel Before the Funnel

Before a contact makes it all the way to your sales funnel, at least in today's modern digital marketing system, it progresses through several stages. It first becomes a contact[28], usually, but not always, by completing a form somewhere on your website. Then the contact becomes a lead by completing any number of other forms on your website as they slowly educate themselves about your offering. Finally, at some pre-defined point, they are considered Marketing Qualified and are passed on to sales. Then a decision has to be made as to whether to accept the lead as a Sales Qualified lead and begin the sales process.

Awareness	Consideration	Decision
2000	1000	7
2250	1100	11
2700	1200	15
3000	1450	25

Table 22. Funnel before the Funnel

What should be clear is that there are normally a lot more leads coming in than what sales is seeing. You should also notice there is overlap at the decision stage. For the purposes of getting to our magic number, however, you can still compute a conversion rate from one stage to the next. You should also be aware that people can enter your funnel at any point in your funnel, regardless of your funnel design intentions (there are exceptions to this rule, of course).

There are also different ways of defining leads and deals and opportunities as well as accounts. These differences can wreak havoc on your ability to manage a sales pipeline when you merge two dissimilar systems, such as Salesforce and SAP or Salesforce and Hubspot.

[28] There are actually two stages prior to this, subscriber and visitor.

Different Strokes for Different Folks

For instance, Hubspot's focus, with regards to a deal, is on the person first. This makes sense, they are, at their core, a marketing company and most of their history has been focused on marketing, which tended to obfuscate the thing a person eventually bought. Once they came out with their CRM tool, that view of the world continued throughout the product.

Salesforce, on the other hand, is an Account Based Model, with Opportunity and then deal, below it. One might argue that this is just semantics, and one would be entirely wrong. Every client I've worked with in this type of environment gets wrapped around the axle trying to understand the differences. Salesforce's model obviously works and makes sense. Then again, so does Hubspot's. The challenge is remembering that difference's in 'point of view' exist as time progresses.

SAP, on the other hand, has a hybrid approach, with a concept called 'Business Partner' as its central organizing entity. It simply recognizes that everybody that has anything to do with the business is in a relationship with the business. It also scales well for large scale business entities. Large meaning a business that may have multiple affiliates and subsidiaries, spread across international boundaries.

When faced with these types of business scenarios, the challenge, even internally for SAP, is to align their sales team with the customer's business. It is not unusual to have many Account Executives working on opportunities involving the same customer spread across the world. Though there may be one primary Account Owner (a sort of super Account Executive), there will also be other AEs trying to coordinate their activities with each of the related AEs.

If you were to do an export of the accounts from within the CRM of say, a Nissan or Boeing, you would see that there are numerous listings of the same customer around the world. Each entity would have one or more AEs assigned as owner. You would also come to see that there are really multiple pipelines running against that one global enterprise. From a Sales Pipeline management perspective, it would be apparent that the idea of a single funnel conversion rate is fantasy. That's why these accounts require a lot of effort to properly model in the CRM so that your sales team can actually pursue all of the different opportunities within one company.

Summary and 10 Lessons Learned

Finding the magic numbers to your sales pipeline takes a bit of sleuthing. When it comes to finding your own sales pipeline magic number, I advise my clients:

- **Make Educated Guesses.** If you don't have data at to start with, make simple assumptions.
- **Install the Analytics Tracking Code.** Getting those numbers quickly is one of the reasons it is so critical to get your analytics tracking code installed and running as soon as possible.
- **Test Your Sales Process with Real Customers.** It is one thing to say you have a 'defined sales process' and quite another to say you have one you've tested with a real customer.
- **Get Sales Pipeline Data You Can Trust.** Data driven organizations do whatever it takes to get the data to figure out what the pipeline conversion rates actually are.
- **Know Your Pipeline Conversion Rates.** Sales Forecasting is going to work a lot better if you know your conversion rates from one stage to the next within your sales pipeline.
- **Always Include the Deal Value.** I always tell my clients to put the deal value in the deal as soon as you create it in the pipeline.
- **Probability Weighted Pipelines Need Deal Values.** Inputting a deal value as soon as you create the deal is required to run a probability weighted pipeline.
- **Accurate Deal Data Lowers Ad-Spend.** One of the little known 'secret' outcomes of having and tracking every deal in the pipeline is that it can drastically improve your ad-spend results, particularly with regards to Google, Facebook and LinkedIn[29].

[29] There are two key ways your CRM system can help you reduce your Ad-Spend and increase your conversion rate. The first is that it can provide immediate feedback to them as to who became a customer. The second is that it can tell them who did not become a customer. This allows the AI (Artificial Intelligence) and ML

- **Keep a Clean Deal Pipeline.** It is as important to tell your CRM system which deals you lost as it is to tell it which deals you won.
- **Leverage Audience Information to Lower Ad-Spend Waste.** A system like Hubspot (and most other CRMs), can use this information to create audiences[30] within these ad platforms, which has huge implications as far as conversion rates.

(Machine Learning) systems these systems employ to improve your outcomes.

[30] ProTip: You want to use customer audiences to: 1. Exclude them from further targeting unless it is an upsell or cross sell scenario 2. Build similar audiences (Google or lookalike audience (Facebook) or matched audiences (LinkedIn) for targeting.

CHAPTER EIGHT

Calculate Conversion Rates

Lonnie Ayers

SIMPLE MATH - POWERFUL PROFITS

Sticking with our numbers for the Sales Funnel, we have three stages in our simple sales funnel. Let's calculate the conversion rates.

Stage One - Inquiry	Stage Two – Demo	Stage Three - Buy
677	34	7
737	54	11
737	40	15
615	56	25

Table 23. Establishing Initial Conversion Rates

Three funnel stages translate into two conversion rates. You could also add a fourth, from Inquiry-to-Buy, but it wouldn't yield any information you could take action on, mostly.

Inquiry to Demo	Demo to Buy
5%	20%
7%	20%
5%	37%
9%	44%

Table 24. Two Conversion Rates From 3 Stages

These numbers can be very difficult to come up with, unless you're on a system like Hubspot, which is designed from the ground up to yield such insight. It's a bit (actually, it's much) harder to extract this information directly from the most common analytics platform, which is Google Analytics.

When a business is first ramping up, it may not have enough sales to make these calculations. Or, in cases like ours, the product we sell often takes a very long time, perhaps 18 months. That's why it's handy to work this funnel backwards.

Summary and 10 Lessons Learned

Conversion rates are a critical piece of information you need to know if you're going to set and achieve growth and revenue goals. But to find yours:

- **Define Your Sales Funnel Stages.** You need to have your sales funnel stages defined
- **Funnel Math is Universal.** It doesn't actually matter what you're selling - the math is the math.
- **Make Reasonable Assumptions Until You Get Real Data.** You could make and use assumptions about your sales funnel to create a reasonable stab at a business plan.
- **Use Reliable Industry Information Sources.** In lieu of actual conversion rate data, use industry benchmarks, which are available from a variety of sources. For example, Hubspot Research[31].
- **Err on the Side of Caution.** If you don't have your own proven conversion rates, be extremely conservative in the use of these rates.
- **Learn How to Play 'What If' With Your Data.** It's best to use sensitivity analysis when using benchmarks and assumptions.
- **Make it Measurable.** Make sure the entry and exit from each sales funnel stage is easily measurable.
- **Instrument Up Your Sales Process.** In order for it to be 'measurable' it must be inspectable. This is possible, with effort, with both ecommerce transactions and other non-ecommerce transaction.
- **Systematize Tracking of Long-Term Sales.** It is simply easier with a system like Hubspot to track what Google refers to as OCT or Offline Conversion Tracking of sales.
- **Watch Out for Double Counting of Sales.** And OCT sale is typically

[31] Hubspot Research https://research.hubspot.com/charts/topic/sales-and-marketing

something selling an hour of a service. Or attending a course. The complicating factor is when there is both a sales pipeline which takes a long time to result in a sale, but the sale is actually closed within an ecommerce system.[32]

[32] As an example, a technical training institute may have a very long sales cycle, yet, the final actual payment is made using a system like woo commerce or Shopify. The deal is closed by the Admissions Officer within the system but the actual transaction is processed via the ecommerce system. From Google's perspective, it is an ecommerce transaction, however, it is actually a combination. This is simply a limit of the current system.

CHAPTER NINE

Worked Sales Funnel Backwards

It's Not That Hard to Do

How do you work a funnel backwards? It's just a question of flipping your funnel upside down. Here again is the simple sales funnel oriented from top-to-bottom or left-to-right

Stage One - Inquiry	Stage Two - Demo	Stage Three - Buy
677	34	7
737	54	11
737	40	15
615	56	2

Table 25. Working the Funnel Backwards 1st Steps

Now let's build our upside-down funnel. Because we want to figure out how many customers we need in order to get to a particular revenue target, we always start from the bottom.

Stage Three - Buy	Conversion Rate	Stage Two - Demo
7	20%	7 ÷.2=35

Table 26. How Many Customers Required

I've just reused the numbers I already had from the chapter on calculating conversion rates. In the following table, I extract the key piece of information I need, which is how many demo's do I have to conduct to make one sale. So just set your "Buy" stage number to 1.

Stage Three - Buy	Conversion Rate	Stage Two - Demo
1	20%	1÷.2=5

Table 27. How Many Demo's Are Needed

Doing the simple funnel reversal math, we find that that I will need to do 5 demos to make one sale. That seems about right, as in my experience, with a highly tuned sales process, I usually see a one-out-of-three win rate. Perhaps this is telling us something useful. Maybe our demo's are ineffective. Or perhaps our lead quality is low. Maybe our offer isn't compelling. In either case, we now have a measurement system and a KPI we can work to change.

Computing the Top of Your Funnel

Keeping in mind that this is a highly simplified yet reasonable sales funnel model, we can now compute how many inquiries it takes to get to 35 demo's, which was the number of demo's I set as a target in the first table in this chapter. Again, doing the funnel reversal math, based on the conversion rates provided, you see that we need to generate 700 inquiries. By convention, we normally are speaking of months as our common unit of time. However, there is no time component in the actual conversion computation. It's just a simple conversion rate formula.

Stage Two - Demo	Conversion Rate	Stage One - Inquiry
35	5%	35÷.05=700

Table 28. How Many Inquiries Needed

Again, as before, the key piece of information I need to know is how many inquiries I need to get in order to get to one demo. Again, setting my numerator to 1 my denominator to 5% and doing the math, I see that I need 20 inquiries to get to one Demo.

Stage Two – Demo	Conversion Rate	Stage One - Inquiry
1	5%	1÷.05=20

Table 29. Inquiries to Demo Required

Now let's build up our business model.

Summary and 10 Lessons Learned

Knowing how many demo's you need to conduct to get to one sale is a key piece of information you'll need to know if you want to set and hit your growth and revenue goals. Knowing the power of one yields the following '**Lessons Learned**':

- **Right Size Your Demo Team Capacity.** Gives you an immediate feel for whether or not your pre-sales demonstration team will be able to help you hit your number.
- **Get The Right Number of People on Your Team.** Gives you an idea of how many people you need to have on your pre-sales demonstration team.
- **Use Time Strategically to Achieve Demo Team Size Sufficiency.** If your product is very complex, you probably have a dispersed demonstration team. Knowing your demo consumption rate gives you some time to assemble the team from disparate locations when needed for complex demonstrations.
- **Get The Right Person for the Job.** When you have very complex demonstrations that must be tailored to specific customer requirements, it pays to get the right specialist.
- **Practice Makes Perfect.** You may have noticed that my computation indicated we would need to deliver five demo's to make a sale, yet I said that it was about right, since I typically saw that it took about three demo's to make a sale, which is quite a difference from the math to the reality. That's because of practice and experience.
- **Invest in Dry Runs - The Harder You Train the Easier It Gets.** Though I cover pre-sales in other publications and blogs, here's what I know from 'lived' experience, ***practice counts***. In fact, before every successful pre-sale demonstration I've run, I've rented hotel rooms and had my team run their demo's to me. That's a practice you should take up. In practical terms, it means for every 'customer facing' delivered pre-sales event, we've actually done it twice. That means in order for me to say three demonstrations sounds about

right in my experience, I am really saying we did six demonstrations, three dry runs; 3 where I played a difficult customer and we delivered 3 'live' events to customers who were, yes, difficult. This is, after all, enterprise class sales. You don't close multi-million-dollar deals without significant efforts.

- **Stand Your Ground on Dry Runs.** Expect push-back from everybody when it comes to pre-sales rehearsals.
- **Watch Your Investment Dollars.** Rehearsals cost a computable amount of money. Bake that into your Cost-of-Sales budget.
- **Do Your Homework Following This Formula.** A good rule of thumb should be very familiar to you - about 3 hours of preparation and demonstration practice time for every hour of demonstration delivered.
- **You Have to Understand Your Customer.** Finally, doing better discovery[33] results in better demonstrations. Don't sap discovery. Canned demonstrations don't win deals.

---

[33] Discovery means the requirements elicitation your team does to gather the 'real' information about customer requirements versus what's in the RFP or other document they provided.

CHAPTER TEN

Leads Required to Make One Sale

FULL FUNNEL REVERSAL CALCULATION

This last part of the funnel reversal calculation is really just the same computation, but from the Inquiry Stage to the Buy Stage. Let's take a look.

Stage Two - Buy	Conversion Rate	Stage One - Inquiry
35	1%	35÷.01=3500

Table 30. Inquiry to Buy Stage

Finally, we can figure out how many inquiries we need to generate to ultimately, generate one sale.

Stage Two - Buy	Conversion Rate	Stage One - Inquiry
1	1%	1÷.01=100

Table 31. Inquiry Required to Make One Sale

If you're running an ecommerce operation, you can visualize the funnel in Google Analytics. However, it is a top-down view designed to tell you how each stage is performing from one stage to the next. Useful for day-to-day performance monitoring, but not for building out your business model. That's why you need to know at least a little excel or have access to a chalk board.

For most businesses, they may not have an ecommerce channel, though they may aspire to get there. For those businesses, you can use what is known as the GCLID (Google Click ID) to tell google which leads became a sale. In essence, it means you can turn long sales cycle type business models into ecommerce type operations. However, it has limits, in particular, their cookies[34] often expire long before you make a sale. That's why you need a system like Hubspot, which has no such limitations.

But this book isn't about Hubspot or even Inbound Marketing and Sales. It is, instead, a 'how to' manual for setting growth and revenue targets, that align with your organization's capabilities. That's why in the next chapter, we're going to play stump-the-dummy a bit. Worse still, I am going to run some word problems by you. Don't worry, the answers are in the book, in particular, this book.

[34] A cookie is a piece of code that Google inserts into a user record when they visit a page. It, along with the Google Analytics tracking code, is used to keep track of which pages a user visited and many other pieces of useful information. All other systems, such as Facebook use similar technology but call them by different names.

Summary and 10 Lessons Learned

Knowing how to reverse your funnel should become second nature to you. By knowing how:

- **You Can Size Your Team Based on Funnel Math.** You have the insight you need to size a high-performance demonstration team.
- **Can Use This Approach to Evaluate Every Stage of Your Funnel.** It provides you with specific, actionable information you can use to evaluate every step of your sales and marketing funnel.
- **Set Numerical Goals.** It provides you with the information you need to set performance goals for each stage of the sales funnel.
- **Can Set Specific Funnel Stage Team Performance Goals.** It also provides you with specific performance goals for each team responsible for each stage of your sales and marketing funnel.
- **You Can Close the Information Loop.** If your business is not an ecommerce type business (or even if it is but your sales cycle takes longer than 90 days), you'll need to get your Google GCLID feedback loop setup and functioning.
- **Can Ensure the Speed You Close Deals Is Within the Time Boundaries of Google.** Google tracks non-ecommerce type sales using GCLIDs and calls them OCT or Offline Conversion Tracking.
- **Convert Your Non-Ecommerce Business into An Ecommerce Business.** The Google OCT approach turns service businesses into virtual ecommerce type businesses. With several significant limitations[35].

---

[35] One of the primary limitations is the length or age of a cookie. Many sales cycles take more than the allotted lifetime of a cookie and thus, a GCLID, which is what Google uses to anonymize and help track sales, 'expires'. Another issue that often arises is that GCLIDs can get dropped, and thus a sale cannot be tracked. It's one of the reasons we've built a WordPress Plugin, GCLID pro, which is available for purchase, which helps solve this often vexing problem. It is an advanced consulting solution we developed to solve GCLID tracking issues.

- **You Can Educate the Google AI.** Once you have your Google OCT tracking set up and get sales feedback going to Google Ads, the Google AI can help you improve your ads results and thus, shorten your sales funnel cycle time.
- **You Can Tell How Long It Takes to Turn a Lead into a Sale.** You'll be able to compute your overall sales cycle time. In other words, you will be able to tell how long, on average it takes a someone to turn into a sale.
- **Set Realistic, Quantifiable, Objective Goals.** As a sales manager, you'll be able to set SMART goals, based on actual-vs.-forecasted data. You can then reward team members who over perform, while 'coaching to success' those who fail to meet standards of clearly defined performance expectations.

CHAPTER ELEVEN

Determine How Many Customers You Must Generate

Road to Target Revenue

Suppose you want to generate $1,000,000 million in annual revenue from your new or even established business. Let's say you sell consulting services. You're a one-man army, and due to job cuts, economic uncertainty, or any number of other factors which you've experienced and have no control over, you find yourself 'on-the-street' and have hung your consulting shingle out.

There's lots of ways to make a million bucks. There are also quite a few ways to price out your services, but most come down to either "by the hour", "by the project" or by the "value delivered". The last is by far the best model, it's also one of the more difficult ones to pull off.

Walk the Model to the Goal

The thing you need to figure out is how much you 'must' make per hour. The second thing you need to know is how many hours are available for you to work in a year. The accepted standard is 2080 hours (per the U.S. government). As it happens, I've built a charge-out rate calculator which you can access here[36].

To use this calculator, there are four components you need to know:

- Your Annual Salary
- Average Utilization Rate
- Your Overhead Rate, in percentage terms
- Desired Profit Margin

Let's say you want to find out what your required charge out rate needs to be if you pay yourself $120,000 a year, have an expected utilization rate of 70%, overhead rate of 30% and finally, a profit margin of 40%. What you'll find is that your 'charge-out' rate is going to have to be about $150 dollars per hour. Though this is certainly a decent amount of money, in reality, it is far from the target of $1,000,000 a year.

---

[36] Charge out rate calculator https://www.sapbwconsulting.com/charge-out-rate-calculator

What will it take to earn your target revenue of $1,000,000 a year using these same parameters? It turns out, you will need to bill out at $1,250 bucks an hour. Though there are certain professions where this can happen, they are few and far between. Lawyers, Doctors, perhaps high-end Business Consultants such as I call myself. In all likelihood, if you want to generate that much revenue, you're going to have to have many people working for you. And keep in mind, I have been referring to Gross Revenue versus Gross Profit.

How Big of a Team?

I am going to keep this model fairly simple and would advise the reader, budget permitting, to run their assumptions by their accountant. But you have to own these figures. Let's make some assumptions to get to a potential model (and we will keep it simple, complexity means excel).

Assumptions:

- Average pay for your team is $120,000 per year
- Average charge-out rate is therefore $150 Per Hour

First pass, but not quite right.

Target Revenue	Hours Per Year	People Needed (FTEs)
$1,000,000	2080	3.21

Table 32. First Pass People Needed Calculation

The process for computing this is just a simple multiplication and then division step. First, multiply the number of hours available per year by the rate per hour $150*2080=$312,000 then $1,000,000/$312,000 = 3.21 Full Time Equivalents or FTEs[37].

Well, that's a good start, but it has one glaring error I left in it, the utilization rate. The charge-out rate is based on, in our case, 70% utilization. This means there are only 2080*.70 = 1456 hours available to work. So now let's rerun the numbers taking this utilization rate into account.

Utilization Rate	Hours Per Year	Maximum Revenue Per Person
70%	1456	$218,400

Table 33. Maximum Revenue Per Person 70% UTE Rate

Now if we rerun the numbers, we see that we need at least 4.58 people billable for the full 1456 hours at this rate to generate our target revenue.

Target Revenue	Hours Per Year	People Needed (FTEs)
$1,000,000	1456	4.58

Table 34. FTE Required to Hit Target Revenue

Business Doesn't Run Itself

You might be wondering who will be running the business while all these people are out earning revenue. More importantly, how many of these people will you need. Though there's no magic number, the Charge-

---

[37] Full Time Equivalents or FTEs is a way to refer to one or more people. You could have 2 people each working for 4 hours a day and you would have 1 FTE

Out rate calculator gives you a hint. You see, overhead is baked into it, and in our example, 30% is sort of the default.

For a small company, in this case, a 5-man consultancy, this number may be smaller, but it will never be less than one person. It may not take one person's full-time attention, but someone has to run accounting and finance, someone has to hire these people, and someone has to run the place, and make sales and most importantly, perform marketing.

There will also be systems required to run this fictional operation. Those will include a financial accounting and management system, a marketing system, and sales system, and perhaps an HR system. In short, the infrastructure required will look like a much larger company. This is much easier to do today with all of the various cloud-based applications available. However, each has an ongoing monthly cost, and none of them run themselves nor do they set themselves up.

Focus on Profit

All of the above scenarios are based on a rather simple focus on revenue. Truly successful companies focus on profit. Without profit, there is only stasis, and soon, rising competition will kill your business. How do all of these numbers change if we focus on making $1,000,000 in annual profit instead of $1,000,000 in revenue. Let's walk through the model (again, highly simplified).

Assumptions are as before as far as salaries go, overhead and profit margins:

Target Profit	Profit % (of Gross Revenue)	Revenue Target
$1,000,000	40%	$2,500,000

Table 35. Target Profit vs. Revenue Target

It now looks like our team is going to have grow, substantially. It still won't be a big company; however, it will now have a requirement for several internal roles that will prove to be full-time. But first, let's take a look at the number of revenue earners you're going to need to bring in that million in profit. Turns out, you will need at a minimum of 12. Or you will need to be selling at a much higher rate. It's also possible you'll need more people if your revenue per employee is much less.

Revenue Target	Maximum Revenue Per Person	People Needed (FTEs)
$2,500,000	$218,400	12 (Rounding Up)

Table 36. Revenue Target vs. FTEs Required

In my experience, at this rate, you will need at least one full-time salesperson (by whatever term you call them), a CPA (can be outsourced but they are not cheap) and access to an HR person or a company that handles this for you. Today, there are a lot of different ways to handle these types of roles, both as full-time employees and as contractors or via service providers.

What about our one-billion-dollar company that we used initially to think through this problem? Is it just a simple matter of adding zero's to the problem? Let's take a look. Are there companies out there that sell only services that this model applies too?

Revenue Target	Maximum Revenue Per Person	People Needed (FTEs)
$1,000,000,000	$218,400	4,579

Table 37. One Billion Revenue vs. FTEs Required

There are many companies that are purely service based. For instance, one of my old employers, whom I worked for in the Middle East, dealt only with people delivering services. In their entire operation, it turned out, they only 'owned' 4 old lease cars that they had bought off their own lease. They had almost 8,000 employees when I worked for them, and as it happens, did about $2,000,000,000 in annual revenue.

Yes, they had more people in what you would classify in overhead roles, but not many. It was a company of systems, running systems on behalf of others, who often times also were mostly companies of systems. But within that universe of people, there were enormous variations in revenue per employee and most importantly, profitability per employee. While some employees might actually bring in $400,000 in revenue and were responsible for a large proportion of the profitability, others were well below the Revenue Per Person figures I've used here.

In fact, from a profitability perspective, it followed a clear Pareto (of 80/20 fame) break-down. Meaning 80% of the profit came from 20% of the employees. Drilling down further and doing a further 80/20 break-down of profitability, it turned out that almost all of the profit was being earned by a small, elite group of overseas defense contractors, of which I was one. As you grow, doing this 80/20 analysis is well worthwhile.

The central message will mostly come out to cull the dogs and feed the heroes. But be careful. Sometimes, lucrative opportunities pop-up in those low profitability but high-volume lines of business.

From Customer to Lead

Now that we know the relatively simple math behind what it takes to earn a million dollars in revenue per year, it's also relatively simple to figure out how many customers we need to generate each year. Again, I will stick to what seems to be a simple business model, selling consulting services.

The first thing you need to understand is that if you're selling consulting, your client expects you to be an expert at solving whatever problem you solve.

The second thing you need to understand is that customers don't have unlimited means (meaning budget or ability to pay) and they have to put a price on every project.

The final critical element to understand is that customers are going to evaluate every project for its ROI or Return on Investment potential. In all likelihood, it means they have numerous projects they are evaluating at any given time and numerous problems they are trying to solve.

What all that boils down to, in practical terms, is that you will have to package your service offering, either in the form of a productized service offering, with clearly defined inputs and outcomes or you will have to package by 'estimated' hours. The last model, batches of hours, may not seem like a great way to differentiate yourself. However, there is an enormous, global market for just such services.

Let's do a simple walk through of how many customers you're going to need to hit your numbers by making some assumptions. You have to start somewhere because nothing has an unlimited price.

Our Pretend Business

Ultra Performance (UP), is a consultancy focused on improving lead generation and sales using Inbound Marketing and Sales for companies with revenues up to $5,000,000,000 and beyond. UP has developed a packaged service offering that includes an analysis of current lead generation acquisition strategies and evaluation of Sales Effectiveness. Most of their customers are in highly competitive industry sectors, such as Aerospace & Defense, Professional Services and Logistics. However, their approach and methodology is just as effective in almost any other industry sector, if executed properly.

UPs productized service package is sold on an annual retainer basis. Their small 4 person teams of Ultra Performance consultants deliver these services over the course of a year, and each consultant makes around $150,000 a year. Their charge-out rate is $187.50 an hour. UP strives to deliver a 10X return to every client. That means that a client who spends $1,000,000 on UP services can expect to receive $10,000,000 in new revenues from their investment. This return usually takes place in year 2 of the project and continues to yield a high ROI for at least 5 years[38], if not longer.

Every project has a Project Manager (PM) in addition to the core 4-person team of Ultra Performance Consultants. The PM is a highly experienced UP consultant and acts as the primary point of contact (POC)

[38] Five years is the traditional convention used when evaluating projected cashflows from a project.

between the client and the team. The PM is engaged at least 20% of the time on the project but is always engaged 100% of the time during the initial phases of the project and during final delivery. However, every single member of the team is expected to interact in a constant fashion with the client team, and each is capable of assuming higher level roles as his career progresses.

Now we have the data to do our simple 'customers required calculation". But first let's recap what we know about UP. They sell a 4 man +20% PM package, which takes a year to deliver. In other words, their service is a one-year long engagement.

Team Revenue	Per Hour	Service Revenue
4 Man Team X 1 Year	$150	$873,600
.20% Project Manager	$275	$80,080
Grand Totals		**$953,680**

Table 38. UP Consulting 4 Man Team Revenue Model

As it happens, if we do a cross-check with our original table where we determined what it would take to generate one million in annual revenue, it comes out real close to our 4.58 FTEs. In this highly simplified model, in which I intentionally created a fictional company with just the right numbers, you only need one customer. As it happens, if I instead target profitability, my model will work, and I will need 3 customers and 12 members of the UP team.

Of course, many might say, how many projects are worth a million dollars a year? Mine are much smaller, lucky to generate $100,000 per year. That's the number we will work with in the next chapter, where we figure out how many leads we're going to need.

Summary and 10 Lessons Learned

Now you know what it takes to generate $1,000,000 in revenue or profits. It seems very simple, doesn't it? Of course, if it was, everybody would be doing it. By knowing how you can:

- **Design to Target Performance.** You can design your business to hit whatever target revenue you want.
- **Design To Achieve a Target Profit Margin.** You can also design your business to achieve whatever profit margin you want to hit.
- **Evaluate The Three Financial Statements To Uncover Useful Insight.** You should realize that you need to have a solid understanding of the three financial statements and how to evaluate them.
- **Understand Your Situation.** You will understand why it is important to have your accountant (or yourself, if you're capable and competent and have the bandwidth) prepare a set of auditable financial statements, every year, if not every month and quarter.
- **Establish a Reasonable Charge-Out Rate.** You've seen how determining your charge-out rate can help you set realistic prices for your goods and services.
- **Sell Higher Value and Get Paid for It.** You should also have realized that what you sell has very little to do with this math. But the more valuable your offering, the easier it is to get your rate.
- **Leverage and Optimize Your Utilization Rate.** You now understand the importance of knowing your utilization rate.
- **Leverage People.** Realize how few people it can take to generate fairly large sums of money.
- **Estimate What an FTE is.** You should realize that using FTEs is an approximation, as no two people are the same in terms of capabilities.
- **Set Your Prices High Enough.** Realize the importance of charging enough to cover your overhead and profit. Without either of those

two elements being covered, you're not going to generate enough cash to sustain or grow your business.

CHAPTER TWELVE

Lead Acquisition

HOW MANY LEADS DO YOU NEED?

When it comes to knowing the number of leads you're going to need to generate, it first is helpful to define[39] what is meant by leads. I am going to be using the definition I use within the world of digital marketing. Within this world, the simplest definition is when someone completes a form on your website and provides a form of contact, at least an email, but ideally, a name and phone number.

You'll notice that, in reality, this isn't much different from what you would get from someone that you met at a networking event or who called in. The biggest difference is in the cost of lead acquisition, which would typically be much higher.

Our Hypothetical Customer

In the previous chapter, I said our typical client at UP Consulting signs-up for a $100,000 project which will take about a year to implement fully. We also said that we are targeting generating $2,500,000 in revenue per year. That means we will need to sign 25 clients this year, or slightly more than 2 a month. In short, this is going to be a challenging year, but we are UP to the challenge.

Marketing though, needs to deliver. How much do they need to deliver? Is there a simple process to work your way backwards so you can set realistic, achievable lead generation goals. As it turns out, there is. It's been around a long time, in one form or another. I call it reversing the funnel.

The first thing you need to know or estimate is your Sales Conversion Rate. Roughly speaking, that is how many leads[40] turn into customers. For illustration purposes, let's say that for every 10 leads you get in, you generate one customer. You would be doing fantastic; a 10% Sales Conversion Rate is nothing to sneeze at. Give yourself a pat on the back.

---

[39] It is helpful, indeed, critical, to define every term used in your business in a data dictionary

[40] I am defining leads here as anyone you have some form of identification on. There are other people who may know your business, which Google calls Users, but who may not have raised their hand yet and made themselves known to you. In reality, you can compute a conversion for these Users as well, but there is less actionable information known about them for you to take, well, action.

However, let's say your numbers are a little closer to reality, and run around 2% to 5%. Let's build us a Reversed Sales Conversion Rate Funnel Table.

Average Order Value (AOV)	Sales Conversion Rate	Leads Required to Generate One Customer
$100,000	.02%	1/.02=50

Table 39. Reversed Sales Conversion Rate Funnel

How does the math on this work? It's like any other word problem, you have to ignore the part that doesn't matter, which in this case is the AOV. And then you just divide 1 customer by your sales conversion rate. In this case, it yields an answer of 50 leads.

Total Leads Required

Since we already laid out the basic data points, I will just set up a table to simplify this for 25 customers.

Total Customers Required	Leads Required to Generate One Customer	Total Leads to be Generated
25	50	50*25=1250

Table 40. Leads Required for 25 Customers

Seems like a reasonable number, given the parameters provided. Works out that you will need to generate a little over 100 leads per month to hit your target. Here's a little secret from someone who has basically been running this scenario on the world's number one Inbound Marketing platform, Hubspot, for 12 years. It is very hard to generate highly qualified leads at this volume for very complex services. It is not impossible, but it will take your marketing team quite a while to start generating these kinds of numbers, especially if you don't have a lot of very high-quality content. In a subsequent book, I am going to help you develop some realistic content production budgetary numbers so you have an idea of what this will cost.

There are, of course, many ways to generate leads. It goes without saying that generally speaking, organic leads, those that find you using a search engine such as Google, will usually convert at much higher rates than those that you acquire through paid methods. Nevertheless, almost all companies that we work with end up spending money on paid advertising, typically Google PPC (Pay-Per-Click), but not exclusively[41].

We have spent, as certified Google Adwords[42] consultants, on behalf of various clients, millions of dollars on Google, Facebook, LinkedIn, Twitter and Pinterest paid ad campaigns. The only thing you have to get right is what you can pay to acquire a customer and still make a profit. In our UP scenario, where each client is worth $100,000 at a minimum, you can actually determine how much you can afford to spend on lead acquisition using Break-Even analysis. For now, let's do some quick analysis on what it will take to generate enough leads using Google PPC.

---

[41] The author is also LinkedIN Ads and Marketing Certified.

[42] The author is a Google Certified Adwords consultant and holds many other Google Certifications

Summary and 10 Lessons Learned

Knowing how many leads marketing must generate is a critical insight. Without it, you can't hold marketing accountable. With it, you can start to ask some smarter questions:

- **Marketing and Sales Must Agree.** Do we have agreement between marketing and sales as to what constitutes a lead?
- **Can We Spend Enough on Lead Acquisition?** Do we know how much we can afford to spend on lead acquisition?
- **Do We Track Lead Acquisition Cost Accurately?** Do we know, down to the penny or very close, what we are currently spending to acquire a lead?
- **Do We Have a Realistic Marketing Budget?** Does your marketing budget make sense overall?
- **Understand How to Reverse Your Funnel.** When you know how to reverse your sales funnel calculations, you have a tool for goal setting.
- **Understand The Limits of Google Analytics.** Though Google Analytics is useful for telling you how many sales each marketing channel generates, it won't tell you what you need to know to make meaningful marketing changes. For that, you need a marketing platform, such as Hubspot.
- **Understand Each Marketing Lead Acquisition Channel.** With your Marketing platform, you can not only determine which channel resulted in how much revenue, you can also tell who specifically came in via each specific channel.
- **Iterate On Buyer Personas and Customer Profiles.** Knowing this, you can do a much better job of developing both your buyer persona and ideal customer profile. This will also drive a need to alter your marketing messaging.
- **All Channels Are NOT Equal.** Some channels are likely to close at much higher rates than others. Generally speaking, organic traffic

is going to convert at a higher rate than paid. But there's a lot of 'it depends' buried in that statement. There is also a need to explore different attribution models. You may find that many channels 'work together' to deliver the lead and cutting back on Ad-Spend will negatively affect your ultimate revenue.

- **Setup Goals in Google Analytics and All Other Systems.** Finally, if you don't have or can't afford a system like Hubspot, (they offer a free version, so not sure what business you're running that can't afford free), keep in mind that within Google Analytics, you need to set up goals[43] in order to measure conversions, and to feed information to Google Ads[44].

[43] One of the useful goals to set up is when someone makes it to your Thank You page after having downloaded a piece of content. You should have their name after this event but if you do not, at least this way, Google knows what type of person converted on that particular page and can further tailor ads to other people who are similar to them.

[44] This exchange of information, even though it is between two Google owned properties, is not automatically setup. You must set it up using the Tools menu on Google Ads.

CHAPTER TWELVE

SETTING A SALES QUOTA

A DELICATE BALANCING ACT

What is a Sales Quota?

In simple terms, a sales quota is the amount of revenue an individual or team must generate per period of time. This can be for a day, month or quarter with an annual grand total goal as well. Sales quotas are set by sales leads. Hitting quota generally results in full payment of an agreed upon performance bonus.

Sales Quotas vs.. Sales Goals

Are sales quotas and sales goals the same thing? Not really. Sales quotas are often part of a series of actions set to help salespeople achieve a certain goal.

Sales quotas are not the same things. Think of a sales quota as steps along the path to achieving a certain goal. For example, let's say sales management of UP consulting has set a Profit Goal of $2,500,000 this year. The first step would be to determine how many sales they need to make to hit the goal.

Next, they would need to determine how many deals each sales person would have to make to achieve the goal. The financial value of those deals would be the salesperson's quota.

The salesperson's compensation plan is almost always tied directly to their sales quota. The plan would include commissions and bonuses.

Sales Quota vs.. Sales Target Metric

Another sales metric that gets mistaken for the sales quota is the sales target. Think of sales targets as set for the team. A sales target will outline how many products or service packages your sales team needs to sell. Sales targets are also time bound. That means, they define how many of what needs to be sold per unit of time.

Having a sales target will help your sales team set specific goals.

Sealing the Deal

Sales reps need transparency. When sales goals and sales targets have been set, sales reps should receive a quota agreement. The agreement should be written and signed by both the sales manager and sales rep. This establishes accountability for hitting goals between the sales rep and their manager. It also keeps managers accountable for accurately rewarding performance.

Sales quotas can be measured in many ways. Smaller companies often use a simple approach, such as so many units a months. Larger companies that sell complex, multi-tiered products and services have more complex quota structures. In these environments, sales people will be held to account for the overall deal value they need to close (i.e., $20,000/month)

Quota structure doesn't end there.

Types of Sales Quotas

- Activity Quota
- Volume Quota
- Profit Quota
- Combination Quota
- Forecast Quota
- Revenue Sales Quota

Here are the five most common types of quotas and examples of each one.

1. Activity Quota

An activity quota requires salespeople — usually BDR[45]s or SDR[46]s — to complete a set number of activities during a period of time, usually one month or quarter. Activities usually include phone calls, follow-up emails, scheduling meetings, and leading demos.

This type of quota is usually assigned to BDRs or SDRs who are part of a larger selling team and aren't responsible for closing actual business. An activity quota ensures they're contributing to the sales organization and providing valuable help to the reps they support.

Activity Sales Quota Example

Sales Rep Doug has a quota of 45 phone calls per month, 84 follow-up emails, and 12 demos each month. These activities are tracked in his CRM and his sales manager can easily see how he's tracking to meet his quota.

2. Volume Quota

Sales reps with volume-based quotas are goaled on the number of units they sell or the total revenue they generate during a specific time period. They're incentivized to sell as many units as they can.

Volume Sales Quota Example

Mickey, with JVN Skates, has a quota to sell 75 pairs of skates each month. He must sell at least 75 pairs to meet his quota. He likely receives a commission on each pair of skates he sells and receives a bonus when he reaches his quota.

3. Profit Quota

This type of quota is based on the gross profit or margin of a dedicated sales team, product/service grouping, or salesperson.

---

[45] BDR - Business Development Representative

[46] SDR - Sales Development Representative

If you're held to a gross margin quota, your number would be calculated by subtracting the cost of goods you sell from the overall revenue. A gross profit quota is calculated by subtracting selling expenses and the cost of goods sold from the final revenue number.

Profit Sales Quota Example

Mickey sells skates to professional skaters. He sells fewer skates at a higher cost than his sales colleague Scott, who sells a larger volume of cheaper skates to recreational skating rinks. Mickey and Scott would both have profit quotas of different amounts.

4. Combination Quota

Some salespeople may have more than one quota. A combination quota might include an activity quota and a profit quota. A combination strategy gives reps a roadmap to success and provides smaller milestones to make their quota more easily attainable.

Combination Sales Quota Example

Mickey's quota includes several activity quotas and a profit quota. He must make 50 calls every month, lead 15 skate demos, and close $2,500 worth of business adjusted for selling expenses and the cost of the skates.

5. Forecast Quota

Forecast quotas are generally assigned to specific sales territories or teams. A forecast quota is calculated based on the historical performance of a region and the revenue goal it must hit.

Forecast Sales Quota Example

Let's say Mickey is the pacific northwest territory rep for JVN Skates and traditionally closes $7,500 in sales during Q4. If his goal is to increase Q4 revenue by 25% this year, his quota would be $9,400 for the quarter.

If Bobby, the rep who's responsible for the southwest sales region, historically closes approximately $4,500 in Q4 revenue, he might have an adjusted quota based on his unique market.

6. Revenue Sales Quota

A simple way to set sales quotas is to use revenue as the goal. That means each salesperson will be responsible for bringing in a specific amount of revenue each month or quarter. While this is a pretty simple sales quota, you can customize this to fit the needs of your business and your sales team. For example, you could incentivize reps to sell certain products by breaking the larger revenue quota down into smaller sections.

Revenue Sales Quota Example

Mickey is responsible for meeting a revenue sales quota of $2,500 per month. Based on historical data, he knows that he needs to sell about 75 $33 skates to meet quota. However, if his sales manager breaks the revenue goal down by product, Mickey can have more flexibility by selling fewer, more expensive skates or targeting low-hanging fruit by selling a larger number of less expensive ones.

Sales Quota Formula

Here's a quick guide on how to calculate sales quotas for your team. Follow this simple formula:

Average number of closed won deals per month x Average contract value = Baseline sales quota

Enemies of a Healthy Sales Quota

Many things can work against your ability to set a realistic sales quota — and your rep's ability to meet it. Get these wrong and you end up missing your goal and probably experiencing high sales team turn-over.

1. Unrealistic Quotas

If you set your quota without considering seasonality or historical rep performance, you're setting your team up for failure.

Similarly, don't adjust quota based on an unexpected renewal or larger than average deal. Raising rep quotas because of a great month can demoralize high-performing reps and stall growth.

Before raising quotas, make sure you have at least three months of data to support the decision — and that it's not just coming from one rep and their specific territory.

If you have one rep or territory consistently beating their number, identify what their success is due to and consider adjusting their quota instead of changing everyone else's.

2. Stress Due to Unrealistic Quotas

The fastest way to see your sales team burn out is by setting unrealistic quotas. As a sales manager, it's your job to know your team's limits and make that argument to your executives when they push for a more top-down approach to quotas.

The average tenure of SDRs has seen a 50% decline from an average of 2.5 years in 2010 to 1.5 years in 2018. Sales is already a volatile profession. Retain top-tier employees by ensuring their quotas are realistic and they have the support to meet and exceed their goals regularly.

3. Commission Caps

Commission caps limit the amount of commission a salesperson can earn. When a rep has hit their commission cap, they are no longer financially rewarded for closing more business. This de-incentivizes reps to push the limits and close more business.

Summary and 10 Lessons Learned

Setting proper sales quotas is a delicate balancing act. There are a number of ways to set them, but no hard and fast rules. However, in my years of enterprise experience, I've participated in a number of quota setting exercises and believe a few 'best practices' are clear. When it comes to setting sales quotas, some of the best practices I've learned include:

- **Size Does Matter.** Smaller organizations with one or two salespeople, especially startups, usually don't have the sales volume to set quotas.
- **Being Big Doesn't Make It Easier.** Larger organizations spend an inordinate amount of time setting quotas.
- **Be Completely Transparent.** The simpler and more transparent you can be with your quotas, the more likely you are to see success.
- **Public Transparency Can Turbocharge Team Performance.** Don't be afraid to set goals and make them public to the whole team. One of the call centers I've worked with had large monitors that reflected Goal vs.. Achievement that completely surrounded the call center. You don't want to have your name appear in Red.
- **Personalize Quotas.** Sales quotas should be set per individual. Much depends on skill and experience.
- **You Need Automation.** If you are going to track activity, you're going to need an automatic system, otherwise known as a CRM.
- **Setting Team Quotas Can Be Problematic.** When you have geographically dispersed sales teams, it is unrealistic to set exactly the same quota for geographically separate unit. Each country and market behaves differently.
- **Each Type of Sales Team Presents Unique Challenges.** Managing an inside sales team is just as challenging quota wise as a field sales team. The more productized your offering is, the more you will see a movement toward more inside sales teams.
- **Competition Can be Good or Bad.** Every sales manager should

constantly be dreaming up new sales contest or incentives, or at least, that seems to be what they think. My experience is that you should tread lightly here. You need to bring in the revenue short-term and long-term.[47]

- **Have an Audit Trail.** Of all the things you must get right when it comes to sales quota, having data that is auditable by the sales rep is both the most challenging and the hardest to provide, especially in complex environments. A side benefit of having systems that help generate auditable data is that you can also spot when things are going wrong and react quickly.

---

[47] One of the unintended side affects of contests and other sales incentives is shifting focus away from very large deals to closable, often very small deals. In my experience, the team on the buyer side of the very large deal will immediately sense the downshifting priority attention and perceive they are no longer important as a potential customer. That's not to say you shouldn't use sales incentives to boost sales, just to keep this behavior in mind, given how much you're likely to have spent already on pre-sales on these large deals which often span years to close.

CHAPTER THIRTEEN

Google Ad Spend

HOW MUCH TO SPEND ON ADS

When it comes to spending money on lead acquisition, there are a lot of different channels you could spend your money on and a lot of different ways to measure results. For our purposes, we are going to focus on Google Pay Per Click (also commonly referred to as PPC) advertising. Why focus on Google?

Near Monopoly Player in the Search Market

When you add up all the different Google properties, which, as of the time of this writing, include Google Search, Google Display, Google Merchant Center, YouTube, Waze, and of course, the Android operating system, it turns out Google holds around 92% of the overall search market. We'll get into the main competitors later, but for now, let's concentrate on this one channel.

When it comes to answering the question of how much you should spend on Google Ads (not just PPC, but all of their properties), it comes down to knowing how much you can afford to spend per lead. What does 'afford to spend per lead' actually mean? At a high-level, it means knowing how much you can spend on PPC and still make a profit. Let's work through examples.

Quick Recap of Our Scenario

We've already determined that we want to focus on generating $2,500,000 in profit for the upcoming year at UP consulting. We've also said that our AOV (Average Order Value) was $100,000 and as a consequence, we will need to close 25 customers, which is just over 2 new customers per month. In reality, you'll need to figure out what the profit is on an order, and then do this math, but the concept remains relevant. We've also said we achieve an overall Sales Conversion rate of .02%, and that means we need 1250 leads to achieve our Sales Goal. Why 1250? Because we need 50 leads to close one customer at a .02% Sales Conversion Rate.

Clicks to Conversion Rate

When you're developing your PPC plan, the bottom line is, you have to know your numbers. When you first start out, all you have is a hypothesis about how your campaigns might convert. Here's a little secret. Google will tell you that if you don't have enough budget to run experiments for at least 3 months, but more likely 6 months, you don't have enough money to do this.

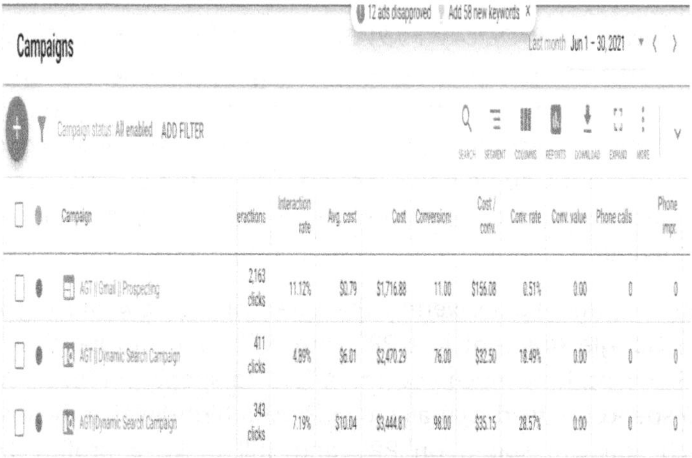

Campaign	eractions	Interaction rate	Avg. cost	Cost	Conversions	Cost / conv.	Conv. rate	Conv. value	Phone calls	Phone impr.
AGT || Gmail || Prospecting	2,163 clicks	11.12%	$0.79	$1,716.88	11.00	$156.08	0.51%	0.00	0	0
AGT || Dynamic Search Campaign	411 clicks	4.89%	$6.01	$2,470.29	76.00	$32.50	18.49%	0.00	0	0
AGT|Dynamic Search Campaign	343 clicks	7.19%	$10.04	$3,444.81	98.00	$35.15	28.57%	0.00	0	0

Figure 1. PPC Campaign View

Real World Example from a Real Customer

The above is a screen shot from within a live Google Ads account that I have been managing for the past 3 years. This is focused on the last month for illustration purposes only. In this case, I have focused on PPC Ad Campaigns that were originally setup by Google themselves. **Pro Tip:** They will do that when your ad spend reaches certain spend thresholds. I've been continuously optimizing ever since. Please pay attention to the bottom two campaigns. We will use them in our analysis.

From the data, you can see that we are paying for clicks, and we're focused on conversions. In this case, a conversion is acquiring a new contact. That is not always the case, as there are many other conversions happening that marketers may want to spend money on. But in this particular client's case, we've been running the campaigns for many years, at about a $3,000,000 ad spend rate per year. The typical sale is fairly low value, at about an initial $230 in the door. However, there are recurring revenue elements of about $5 a month that get tacked on, and about an 80% customer retention rate across a five-year period. Organic traffic is virtually non-existent as is the case with all other channels.

Number of clicks	Cost Per Click	Cost Per Conversion
343	$10.04	$35.15

Table 41. Cost Per Conversion Based on Clicks

Another way of stating the above is that it is costing $35.15 to acquire one lead. Many people define a lead in many different ways. They also change the name of a lead as it progresses through their sales funnel. You'll hear such terms as MQL[48], SQL[49] and Opportunity bandied about. What's important to know now is how many leads do you need to work to get to a sale.

As before, it is just a conversion rate question. You can either do your conversion rate math at $230, the initial value, or at $290, the 1st year revenue value, or you can get a little more adventuresome and go for the 5-year value times 80%. I can tell you from experience, all customers initially focus on the initial value. This is true up until the value of the sale exceeds a certain threshold, starting at $500. The higher the value of the sale, the more you should be willing to spend on a lead.

In our example scenario, the absolute maximum you should be willing to spend is found by dividing the value of the sale by the cost per conversion. All you have to do is divide $230 by $35.15 which is 6.54 leads or a maximum of $230 you should be willing to spend on leads to make a

[48] MQL, Marketing Qualified Lead. This is an entirely arbitrary call as to when marketing considers a lead as marketing qualified.

[49] SQL, Sales Qualified Lead. This is also an arbitrary call on the part of sales. It simply means the lead is, in the eyes of Sales, ready to be worked. That can mean many things to different organizations.

sale. However, you can't spend 100% of the value of a sale on lead acquisition. You have many other costs that go into making the sale, such as overhead, IT systems and other elements that go into COGS or Cost of Goods Sold. Ultimately, you need to know what your break-even rate is. In our case, this customer has determined that their break-even rate is $140, which means they need to use only 3.98 leads to generate a customer, or they will lose money, at least in the first year.

Now that we know their basic conversion rate numbers, including the cost per lead, the number of leads they need to generate a sale (actually, the maximum number of leads they can consume to make a sale is what we know, using their break-even analysis), we can move to the next step in the process.

In our case, this is a fairly high-volume client, and they want the paid lead generation channel to deliver 2,000 sales a month or 24,000 sales a year. Doing the math, that means they want PPC to deliver 24,000 * $230 = $5,520,000 in net new annual revenue. Their overall annual sales revenue goal is about $24,000,000, much of which is delivered from previous customers.

Now we have all we need to know to develop our annual PPC budget. We know we need to deliver 24,000 new customers this year and that we can 'consume' a maximum of 3.98 leads per customer. That's 95,520 leads at a cost per lead of $35.15 which translates into an annual PPC budget of $3,357,844 plus whatever they have to pay for ad-spend management, which will be somewhere between 15% and 20% of annual ad-spend (I typically charge a percent of ad-spend plus a commission on every sale, as long as I have total transparency into the customer's complete sales process). Though that sounds like a lot, even in the first year, this ad-spend is going to yield $2,162,156 in net revenue after expenses, and then yield even more substantial out year cash flows, given the 80% customer retention rate.

Of course, this client also has substantial cross-sell and upsell opportunities embedded in their business model, which makes this scenario look even more profitable. Due to the integrated nature of their digital presence, and Google's tracking abilities, we are able to deliver pinpoint messaging to these existing customers as they explore other needs they may have.

High Ticket Sales

The previous example is handy, first off, because it is based on a real-world example from this author's actual business and secondarily, because it allows me to contrast low dollar value sales planning with high-ticket sales.

What are high-ticket sales? Fundamentally, they are above about the $2,000 range and up. However, I want to focus on the much higher value sales. In particular, on the SAP space. Why SAP? Because that is where this author has spent over 20 years (and counting) driving very large dollar sales and because the numbers are so different from a small dollar, high volume sales operation.

What are we talking about when we talk about SAP sales. First off, please keep in mind that these numbers change over time, and what I am about to share with you are probably going to be very different from your own experience.

The typical AOV of an SAP sale is about $100,000, just for the software license, this is way down from the figure in 1999, the first year I worked within the SAP space. On top of that, you will also see 22% in annual maintenance fees tacked on, for software support. As a rule of thumb, somewhere between 1 and 10 dollars in implementation fees will eventually be spent on implementing the system. The larger the system, the closer it will be to $10. Prior to the emergence of the cloud, the rule-of-thumb for hardware and networking infrastructure was about 20% of the overall project cost. Let's use, given my experience, $3 dollars for every one dollar in license fee as an average implementation cost.

Now we have some rough numbers to estimate the cost of a small, yet fairly average new SAP implementation. Summing them up, we get $100,000 (License) + $300,000 (Implementation), +$22,000 (1st year support) + $80,000 (Hardware) for a total 1st year cost of = $502,000

However, these are averages. In my role as a Quota Carrying SAP Industry Principal, I saw a very different set of numbers which I'll walk you through now.

My annual Net New License Sale Quota was about $4,400,000. My very generous bonus was based on hitting this number. I normally achieved 500% of bonus, meaning I sold $22,000,000 in net new license sales annually.

I didn't make those numbers by selling 220 new customers at an AOV $100,000. Instead, I concentrated on very large deals, whose software alone usually cost in the range of $1,000,000 and up, sometimes, way above that. Why tell you all these sales numbers when we are trying to set a Google Ad Spend Budget? We'll get there, but let's figure out the numbers on a single deal, using some of the inputs I gave you before.

In this scenario, let's assume I have a deal in the works whose license value is $1,000,000. Due to its nature, larger deals typically drive the implementation cost way up, and in this case, it is working out that it will take $10 dollars of implementation for every one dollar in sales. That means our implementation cost is going to come in at $10,000,000. As before, support will be 22% of license cost, annually, or $220,000. Hardware and networking will also be assumed to run about 20% overall or $2,200,000. There will be other cost, such as training and change management, but those are baked into the implementation fees. The overall initial 1st year cost is therefore estimated at $12,520,000.

To be realistic here, the only number that actually counts, when it comes to deciding upon how much you're willing to spend on PPC, is the SAP License. Why is that? And what actually happens? Well, if you're SAP, the most profitable part of this transaction, at least initially, is the SAP software license sale. After the initial sale is made, much of the subsequent revenue comes along automatically, such as support. If they deliver the implementation, they do OK on that as well, but it, too, gets attached to the original sale. If a partner delivers it, the partner takes those dollars, mostly.

So, we have a number to work with, $1,000,000. As before, the answer comes down to your break-even rate. I can tell you from experience that these deals are NEVER made by just one person, and that every lead, on average, can cost $100 to much, much more to acquire. Based on my experience, somewhere between 9 and 13 people were typically part of the customer's software solution selection committee. Let's go with the worst-case scenario, 13. And let's assume you need to 'convert' all 13 of them at a cost of $100 each, or $1,300 in total. Now it should become obvious that you're not really going to worry too much about spending $1,300 on lead acquisition for a million-dollar sale. Unfortunately, it isn't quite that simple. That's because of the sales conversion rate and everything else it takes to make a sale.

In an operation like SAP (or really, any other large customer), they have very large, complex marketing operations teams that generate leads from a variety of sources, of which Google PPC is just one, but yet, critical channel. The reason it is a critical channel is because of how many of the more traditional channels have continuously decreased in performance.

Quick War Story

I once needed to get a lead, meaning make contact with, the CEO of a soon to be built brand new railroad in the UAE. I spent $50,000 on a booth at a railway marketing event in Abu Dhabi, and acquired a total of 9 leads, at a cost per lead of $5,500! As it happens, we did make contact, and that lead turned into yet another SAP Railway Customer, probably the only new one in 30 years. However, it took many years to close that deal, and there is no way to justify that cost per lead given the cash flow timelines, even among a giant like SAP.

As a result, when I struck out on my own as a SAP focused consultancy, the SAP marketing director pushed me to use Inbound Marketing as a primary lead generation strategy. SAP now uses this strategy extensively, but at the time, it was considered very radical. Instead, they relied heavily upon a complex multi-channel marketing mix, including massive spending on Google PPC, which is still the case to this day.

Why Your Sales Conversion Rate Drives Your Spend Decisions

Ok, now that you know a bit more about my background in SAP sales, let's circle back to figuring out how much to spend on Google PPC if you're selling high-ticket items.

Given that I've already told you it is going to take 13 leads at a cost of $1,300 for a $1,000,000 sales opportunity, what you really need to know is how many $1,000,000 sales opportunities you're going to need to make one sale. Stated another way, you're going to need to know your Sales Conversion Rate.

The math here remains the same, it's just the sales conversion funnel. Let's restate it with a $1,000,000 AOV and an optimistic .05% Sales Conversion Rate.

AOV	Sales Conversion Rate	Leads Required to Generate One Customer
$1,000,000	.05	1/.05=20

Table 42. Leads Required to Generate One Customer

Now, some of you may wonder if .05% is a realistic conversion rate. Some will say, ha, I close 50% of my deals, while others will say, I don't close but one percent of mine. Let me clarify and set expectations because this one is a little more complicated.

You see, at these deal values, .05% may well be an overall valid Sales Conversion Rate, if you measure from the very beginning of the lead acquisition pipeline all the way through, which you should. But for a guy like me, in the above scenario, I was working in a subset of the acquisition pipeline, and the conversion rate was closer to 33%. That rate, however, is a measurement of something completely different from your overall Sales Conversion Rate.

You see, for high-ticket sales, such as SAP software, there is an entire lead acquisition machine that works before the sales team engages with a lead. The sales pipeline is also very complex, consisting of something like 13 steps (which are really only visible to the sales team), each of which has a conversion rate from one stage to the next. This is called a Probability Weighted Sales Pipeline. When you have a marketing system which is generating a huge number of leads, the basic 'secret to success' is too quickly qualify out leads. The deeper they are in your Sales Pipeline, the more 'sunk cost' you will have in them, and thus, the harder it is to 'let them go'.

Another Quick and Painful War Story

Before I tell you about a recent loss, I should let you know that I don't necessarily ascribe to the quick 'qualify them out' philosophy but will do it when I am 'fairly certain' the customer is not going to buy, at least at this time.

That said, as a sales manager, with deep expertise in the entire SAP approach to Value Selling, I have substantial experience with delivering SAP

Pre-Sales demonstrations[50]. However, even I sometimes allow myself to be sucked into the heat of the moment vortex of 'this is a done deal' often sold by Account Executives[51]. But in this little vignette, in which I was working with an SAP Partner as their Director of Sales and Marketing, who was supporting SAP in one of their mega sales opportunities, I allowed a step to be skipped, which I knew was critical.

In this deal, which we ultimately lost, we, along with SAP and a number of other partners, performed the normal 'Discovery' process and ultimately delivered what turned out to be a 3 week long pre-sales demonstration to about 250 customer representatives from across the 17 Billion USD corporate group. The demonstration went, at best, 'so so'. We didn't really run into any technical glitches during the demonstrations, nor did we actually manage to demonstrate the technical differentiators. I would give us a B to B+ on this one. And I graded on a curve.

What we did not do, despite my repeated questioning about it, was deliver the business case, or in SAP speak, the Value Case. Why is that such a huge deal, and how is it relevant to determine what you should spend on Google PPC or any other lead acquisition channel? Let me explain.

You see, almost 20 plus years ago, SAP realized that they had to 'up their game' when it came to delivering value. Though their solution was 'state-of-the-art' when it was released more than 40 years ago, and they had and continue to continuously roll out new functionality, there are many other able competitors in their market space. In particular, they (SAP) face off against Oracle and Microsoft in every single deal. They also have about 58[52] other competitors, just in the ERP space. Beyond the ERP space, they have hundreds of other products, all of which face their own set of competitors, and all of whom are just as determined to win. In short, it is a hellscape of competitors.

While competitors are an ever-present threat, customers themselves employ a wide variety of strategies to both make the right software solution selection, but most critically, a whole host of techniques to drive all software vendors into a features and functions battle, ultimately resulting in

---

[50] Over a period of 21 years, I estimate that I have participated in over 200 SAP Pre-Sales Demonstrations

[51] Account Executives or AEs is just what SAP calls Salespeople

[52] That was how many I had identified in 2011 and for which I had competitive battle cards at the ready

a 'how low will you go' scenario. One of those strategies is to employ a 3rd party to develop their requirements and ultimately, conduct a software selection process.

This is the scenario we found ourselves in on this misadventure. The reason the customer wants you in a features and functions battle is to drive down your price. The counter-measure is to employ and deploy appropriate business case development tools and methodologies from the beginning until the end of the process. SAP has developed an entire suite of tools and methodologies for this exact purpose, roughly called Value Lifecycle Management.

System Integrators, who are often the 3rd Party evaluators hired by customers to conduct these exercises, will typically do everything in their power to prevent companies like SAP from employing these tools. The proper response, proven time and time again, is to qualify the lead out of the Sales Pipeline and move on. In this case, we did not do that. Like old hound dogs, we could smell blood in the water.

Ultimately, after we all had invested a substantial amount of time and money in preparing and delivering this demonstration, SAP came to the conclusion that the client was not qualified and fired them. You read that right, SAP fired a client. A post forensic sales process audit revealed the lack of a business case, along with a misaligned IT deployment strategy on the part of the customer, made the client unsuitable for SAP.

But why did we get so far into the sales process before we pulled the ripcord and bailed out? It's very common, actually. There is an inherent conflict between the need for the AE to make quota and the need for the entire extended SAP team, including partners, to do the Value Engineering work, which often takes a lot of time to do properly. You see, the AE is managing many opportunities, and his Sales Manager is asking him or her to commit on the Sales Forecast. This sales forecast ultimately gets rolled all the way up to the top level of the business and forms part of what the CEO briefs financial analysts on. You will often hear the term 'guidance' bandied about and these forecasts form part of that very public communication.

Conducting a full-fledged Value Engineering exercise is not free. If SAP can, they will get the customer to pay for it. If they cannot get the customer to pay for it, but the Industry Principal believes it is a worthwhile opportunity, then the cost will be covered by an Internal Order. That's a fancy way of saying cost will be booked to the cost of sales for that opportunity.

Here's where you start to consider how much to spend on all aspects of the opportunity, including PPC. You see, when the opportunity pipeline is full to overflowing, you can be very picky about which deals you decide to pursue. By you, I mean Sales.

Within the Marketing and Sales world, at least in the ideal model envisioned by everybody, Marketing would be tasked with delivering a certain number of MQL or Marketing Qualified Leads to Sales. Sales would be tasked with reviewing and accepting those MQLs and turning them into SQL or Sales Qualified Leads within a given time frame. Then Sales would commit to and actually work to convert those leads into Opportunities, and ultimately sales, within a set period of time. All of this would fall under the rubric of a Service Level Agreement or SLA established between marketing and sales.

It sounds nice. And for many companies, this model is perfectly doable. But I find that as the deal value expands, and thus complexity expands, the process flow doesn't quite work as clearly as this. Instead, marketing delivers some of the leads, and they also perform many other functions, including putting on events, acquiring customer testimonials (very time consuming and difficult) and many other activities. That is what they have budget and manning for.

What actually happens is that, at least from the Industry Principal perspective, is that although a few leads come in from PPC, many more come in through networking. And this requires your physical presence and activity. The challenge is to leverage both channels while obtaining enough budget to fully exploit both channels.

In the case of a new lead coming in through PPC, then the most probable next activity will be further opportunity development using networking and other strategies to expand the number and depth of contacts within the opportunity.

That is one of the reasons that marketing is expensive, getting more expensive and is unlikely to get less expensive. More competition, more savvy customers, and when it comes to SAP, not too many 'whales' left in the ocean. However, they still manage to close, according to SAPInsider, between 8000 and 9000 new customers a year. These will look like a pyramid if you map them out by deal size. What is not divulged is that much of their growth comes from expanding their existing footprint among their install base, meaning existing customers.

Where Partners Dare to Tread

It is this last activity where I've found myself playing a key role. You see, there are something like 20,000 SAP partners out there. Most of them are tiny, and even the very large ones, don't focus just on SAP. The small ones drive a huge proportion of the footprint expansion that SAP enjoys. This typically happens because the partner is hyper specialized, with deep expertise in one particular part of SAP.

In fact, these specialized partners will often become who AEs turn to first to help them pursue a deal. Why? Because they can't get the same help within SAP. In many cases, this expertise exist only among one or two specialized partners in the entire world. In others, there will be areas of expertise that are only handled by a small number of experts, who SAP did not and cannot replicate. For instance, in my area of expertise - MRO or Maintenance, Repair and Overhaul of aircraft, the expertise is thin on the ground. In the entire SAP eco-system, there are perhaps 5 or 10 people who work on MRO and of those, I know of only one who ever actually turned a wrench on a jet, me.

I've run across other 'super' specialist in many areas. One in particular, struck me as particularly key to SAP, a guy who specialized in hedge management. I had to bring him in on several airline deals which involved Fuel Management and Hedging and he always seemed to be on a ski slope somewhere or another. He was doing something right.

The central lesson here, at least for consultants, is that if you can, specialize. If you cannot, then develop your network of specialized consultants whom you can call upon. You can upsell them, and you'll be helping clients solve an intractable problem - getting the right expertise on the problem, whatever that problem might be.

There is one other lesson embedded in this: SAP does not often create these specialists. There's a lot of reasons for this, most of which a potential SAP customer is never aware of. But the extreme paucity of SAP Consultants in a particular specialty is sure indicator that the particular solution is not well supported by SAP. This may or may not be a problem. If they have a solution that meets your need, then it is usually technically better to go with that as SAP is committing to support it.

If you're one of these super specialists, however, there's probably a huge opportunity to use Google PPC to uncover hard-to-find consulting opportunities.

Offline Conversion Tracking (OCT) - Key to Sales Success

For both hard to find specialist, whether SAP or any other flavor, or any other firm whose sales cycle takes a while[53], you need to be aware of a concept called Offline Conversion Tracking (OCT). What is OCT? It is a Google's current term for tracking non-ecommerce type sales. Going a little deeper, it means tracking sales that take a long time. Most complex, high-ticket sales fall into this category. For instance, even the simplest of SAP sales, even if for a small project that will involve no more than 3 consultants and take perhaps a year or less, will take a long time to close.

For these types of sales, a lead, when initially generated by a Google PPC ad click, will generate what is known as a GCLID, or Google Click ID. What is it and what purpose does it serve? What are some of the limitations for using it and what must you do to take advantage of it?

The GCLID is a long string of characters and numbers that are appended to every click they send your way from a PPC ad. It gets appended to the URL of the page where the click is directed. Its purpose is manyfold. The first is that it, in accordance with Google's ever tightening data privacy policies, anonymizes a user's ID. They don't want you to be able to connect a particular user to a particular ad nor sale. Admittedly, that's not very helpful to you, the advertiser. That's why you need a system like Hubspot, where you can capture leads.

So now you know that not only is Google planting a tracking pixel on your website from every visitor, whether driven by an Ad or organically or any other channel, Google is tagging all the paid clicks with a GCLID. This means that if someone uses Google Search and comes to your site from this Google search, they get cookied.

---

[53] In the case of Google PPC, 90 days is the limit of how long a sales cycle can take, as after this length of time, the GCLID expires.

GCLIDs Drive Higher Sales

The other reason for appending this GCLID to the click, apart from separating and anonymizing user data, is so that you can tell Google which click turned into a Sale. Why would they care? Because Google Ads uses Machine Learning to continuously adjust who they show your ads too. Though this book isn't going to be a complete treatment on Google AdWords, (I will probably end up writing another book on the subject), this particular aspect is critical for businesses whose sales cycle is long, yet it is also limited by several technical considerations that few are aware of.

If you have the system in place to tell Google which GCLID turned into a sale, and how much that sale was worth to you, Google can do a much more effective job of finding the right person to show your ad to.

What happens if you don't have this in place? When you set up an ad campaign in Google Ads, you also have to tell it what conversion action you want the campaign to focus on. You can actually focus on many, but for maximum effectiveness, you want to be very precise with this setting. Many clients will set up a campaign and focus it on lead generation, versus sales. Like magic, if all the pieces are in place, you will start to generate leads from these campaigns. However, you may or may not generate many sales, and you certainly won't generate the maximum number of sales.

Set Campaign Level Objective to Sales

For a campaign that is focused on sales, you want to set the campaign objective[54] to Sales. To make this work for you, you have to set up closed loop GCLID feedback. That means that you have to have the systems in place to tell Google which click (with a GCLID) became a sale. You can do this in a variety of ways, but they boil down to two distinct ways: manually or automatically. If you do not have the systems in place to automatically tell Google which GCLID became a customer, such as you can do with Salesforce or Hubspot, then you have to do it via manual file upload. This can also be automated, and some clients use Zapier[55] successfully to populate Google Sheets with updated sales information.

---

[54] Setting this influences the recommendations Google provides to improve your ads.

[55] Zapier is a cloud-based software integration platform. It cost so much a Zap plus a regular subscription to use.

With this feedback loop in place, you can start to tell Google what's working. Couple of caveats here. First, you need at least 30 sales events for the machine learning system to have enough information to start dialing in the results for you. More sales provides better data for the algorithm to chew on. The other one is that you need to tell Google the truth when it comes to sales figures. That means, if a sale is worth a million dollars, telling it is only worth $1 is just shooting yourself in the foot. You see, you're bidding for clicks in a competitive environment, and Google is in the business of selling clicks to the highest bidder. Though you might be willing to pay the most money for a click, you may be up against a competitor who also has a higher quality score[56] (awarded by Google) and they will win against you.

You also have many options when it comes to reporting Sales Revenue to Google, and they are rapidly evolving their bid strategies. The most powerful strategy is to bid on profitability. This comes down to telling Google how much profit a sale is generating and letting it optimize on that.

All Good GCLIDs Eventually Die

Now for the bad news. When it comes to using OCT, even if you have all the technical elements in place to use it, a GCLID comes with an expiration date, currently 90 days. Practically speaking, that means when you go to upload sales to Google that happened more than 90 days ago, you will get an error that says this GCLID is too old. There is currently no way to go past this time barrier. Google is working on ways to accommodate longer sales cycles, but for now, 90 days is the hard limit.

Why do they die? Though I can't claim to speak for Google, they have explained to me that it comes down to what happens to the click as a user moves through time. The first problem is that people have to periodically clear their cache to resolve problems. Though there are technically savvy users who know how to clear only a portion of them, almost all of them simply clear it all. Once that's done, the chain of data custody is lost between the user (really, between the device the user was using at the time of the click) and the GCLID.

The second cause they've mentioned is that people change devices.

---

[56] Google grades your ads and your landing page and awards it a quality score, which directly affects how your ad performs.

Yes, Google has the ability to perform cross-device tracking very effectively. However, when your actual device dies, or you get a new iPhone, that also breaks the chain.

What My Google Rep Told Me

Now I am going to reveal something that few, if any advertiser, is ever told, as conveyed to me by a very good Google Account Representative. Don't waste money on Google Ads unless you can close the loop on sales. That's right, if you can't close the sales loop using their GCLID technology, one of their best said he wouldn't spend money on Google Ads. If you're also driving Inbound Calls using Google Ads, as many of my clients do, he went further and said, unless you have Call Tracking set up to include closing the loop on sales, don't waste your money.

What that all translates into is, if you can't set up closed-loop analytics, from the first click all the way back through to sending the GCLID to Google with sales information, you are wasting your money. Without this closed loop, even if you thoroughly know and understand the PPC formulas I've walked you through in this and previous chapters, you're going to spend far more on ads than you should. That's why we first see whether or if your loop is closed or closable.

Multiple Elements to Evaluate

When you get right down to it, there are multiple elements to get right with something that is sold as simple to do and highly effective at driving sales. Those elements are:

- Having a Target Revenue Goal
- Knowing your Click-to-Conversion Rate
- Having a closed-loop system

But there is one other critical element that must be in place to make your Google PPC ad-spend work. Your offer. The key to getting your Google Ads to work for you is for your ads to be completely aligned with your offer - what the prospective buyer sees on the page you're sending them to.

Always Work on Your Offer

Why mention your offer in a chapter on how much to spend on Google PPC? Experience. That's right, experience tells me that if you don't have an offer that is competitive, with or without Google PPC, you won't win. But if you do have a great offer, you still have to make sure the keywords you're targeting, the ads you're writing, and the offer are tightly aligned.

Google provides a very helpful metric here, a quality score. It actually provides two different quality scores, one that is focused on your overall quality score, with regards to the keywords you're targeting, and the other one is a quality score with regards to the landing page. The second one is somewhat subjective, but is done by objectively measuring various elements as defined by Google.

From the standpoint of getting the most bang for your advertising buck, you need to make sure you're targeting a Quality Score of between 7 and 10 on both elements. I generally periodically, at least weekly on accounts I am managing, go through my target keywords and pause those keywords that are rated below these numbers. That's not a 'set in concrete' rule. I also take a look at the conversion rate I am getting, and if I am getting an acceptable conversion rate, even though Google is giving it a low-quality score, I may let it continue to run.

During these periodic evaluations, I will generally also look to create a better offer page, better ads aligned to these offer pages and a narrow, better aligned set of keywords. It's a never-ending task. But unless and until I can get to the data as to what it will take click wise to get a sale, it is unavoidable. You would be shocked at how many accounts I've opened the hood on and found that they are spending massive amounts of money, but don't even have conversion tracking set up, let alone knowing what their sales conversion rate is from PPC. That's one of the key reasons Google is so big - many advertisers are spending wastefully. That's why you have to figure out whether Google is the right channel, or some other channel or approach is the right way. In short, you need data to spend wisely.

Summary and 10 Lessons Learned

Google PPC can be a highly effective sales channel. However, it requires a number of key elements to be set up correctly to avoid wasteful Ad-Spend. What you must keep in mind when it comes to this channel:

- **A Good Offer is a Prerequisite.** A bad offer will outweigh any good PPC practices you employ.
- **Report Actual Sales and Sales Values to Google.** Setting up closed-loop reporting that includes actual sales using the Google GCLID is a prerequisite for achieving acceptable performance from your Ad-Spend.
- **GCLIDs have an expiration date.** Always monitor your upload status if you do it via flat file uploads.
- **Ensure All Digital Platforms Are Fully Integrated.** If you have a system that integrates with Google Ads, like Hubspot or Salesforce, then make sure you have this setup and running.
- **Budget For Sufficient Testing.** You need to have enough budget to run ad test for at least three months, six is better.
- **Implement Call Analytics.** If you run Google's powerful call-only ads, you need to have call analytics set up or don't run them.
- **Measure Your Sales Conversion Rate.** You must know and continuously measure your overall sales conversion rate for this channel.
- **Use Google Ads For High-Ticket Sales and Low Ticket.** High-ticket sales are possible using Google Ads, but generally, are best employed at the top of the funnel to get the initial contact and interest established.
- **Google Grades Your Ad and Your Webpage.** Your quality score should be closely monitored and you should constantly strive to improve it.
- **Focus on Value Based Bidding.** Your ultimate conversion strategy should be to get to campaigns that are bidding on value.

CHAPTER FOURTEEN

Ad Spend Strategy Across Platforms

SPEND YOUR AD DOLLARS WISELY

Though Google currently absolutely dominates the paid digital advertising landscape, with about 92% of the search market as of 2021, there are many other channels to explore and potentially spend your advertising dollars on. Let's do a high-level overview of the top 3 I use with most clients. At the end, I will provide a brief overview of the 'other' channels, which may be perfect for some businesses, but which are special use cases and beyond the scope of the book

Facebook - Yes, People Still Use It

With over 2 billion users globally, Facebook is hard to ignore. Though I personally believe they have made a major strategic blunder by getting involved in politics, they still provide a platform that is very powerful for savvy advertisers.

The advice here is the same as for all other advertising channels. Know your numbers. Like Google, LinkedIn, Twitter and Pinterest, Facebook provides extreme targeting using what is known as the Facebook Pixel, a line of tracking code that you insert on your site. Unfortunately, there is also quite a bit of code to implement to track events, though there are plenty of experts to set you up, such as myself, so that you can track specific events on your website, for example, someone attending an event on your website or visiting a pricing page.

Once you have all of your tracking codes set up on your website, you'll also need to have set up Facebook Business Manager, which is essentially a poorly designed integrated development environment for ads. Like all software tools, it is undergoing constant updates, so there is little point in me providing specific instruction on using it in this book.

The single most powerful feature of Facebook Advertising is the targeting abilities and the ability to create 'look-alike' audiences. First off, there are well over 200 different targeting choices available, and when you have the budget, these can be combined into well over 5000 combinations of parameters that target almost any interest or activity. Want someone who is into weightlifting and gourmet cooking. You can target it.

However, the secret sauce to Facebook advertising success is to create audiences from people who actually became customers from Facebook. Once you get enough Sales Conversions from the Facebook Channel, you can create a look-alike audience of approximately 2,200,000 people (that's just the average number it breaks look-alike audiences into), and you can create multiple look-alike audiences with varying degrees of probability of closing. You can also upload lists of customers who did not come from Facebook, and let it do a behind the scenes audience matching exercise, which can be very helpful in the beginning of your Facebook advertising effort.

Once you have enough of these audiences (actually, before), you can design a set of ads that target each element of your funnel. That means there are ad types for building brand awareness, building engagement and finally, closing a deal. You should know there is a Facebook Blueprint Certification available for free (the certification has a cost, the course is free) from Facebook. I've completed all of their available training, and they are always adding new courses, especially when they acquire new properties, like Facebook Messenger or Instagram.

Facebook Messenger

Facebook Messenger is one of the most used if not the most used apps on the internet. There is a whole world of ecommerce going on within Facebook Messenger. There's also a wide range of messenger based chatbot builder tools out there, one of which, Manychat, I have used to build dozens of highly performing chatbots.

Like every other social media channel, Facebook Messenger and ManyChat can become a full-time career. But I just want to briefly mention a few key activities you should be exploring: Lead Generation, Contests, Customer Service and Quizzes (which can be a flavor of contest). Though the tool is constantly improving, it still takes considerable time, effort, and skill to dream up uses for it, build the chatbot and deploy and integrate it. Yet it can be powerfully worth it, if done right. I've seen it generate thousands of new leads overnight from a single sponsored contest bot, which brought in 8 figures.

Decide How Much to Spend Based on the Data

Like all paid channels, at least when using digital technology, you can and must know your Sales Conversion Rate and your ROI using your real expense numbers and real revenue numbers. The formula for this is the same as that provided in the Google section. Then it comes down to comparing the two channels (or however many channels you are advertising on). If Google is outperforming Facebook, then kill Facebook. Don't harbor any emotional connection to it or any other channel. Just kill it. When it comes to paid advertising, your budget is almost always going to be limited, and it cost time and money to run ads on both channels. The smart strategy is to pick winners and cull the dogs.

LinkedIn For the B2B Market Win

Many, if not most users of LinkedIn, think of it as a job board, which was its purpose when it was founded. I was one of the first 20,000 people to sign-up for it, and I certainly thought it was designed for that purpose. It has provided me many job opportunities over the years, so I can't complain about the results.

But, over the years, it became a giant sales platform, containing information that would get you sued in an earlier time, just for divulging it. Namely, it contains the corporate hierarchy of almost every business on the planet. Though it is primarily a platform for job seekers and recruiters (and recruitment software is still one of its primary revenue sources) it is also a highly effective, but much more expensive, advertising platform for B2B type businesses, and possibly B2C type businesses.

Like the other platforms, such as Google and Facebook, LinkedIn also has a tracking code that must be placed on your site, called an Insight Tag. Unlike the other platforms, you also have to place a different piece of tracking software on the 'backside' of your conversion page if you want to create leads on your website using your LinkedIn lead ads campaigns (there are many types of LinkedIn ads with different goals). The other huge difference is the audience manager on LinkedIn. You can, of course, create very highly targeted ads which you can aim at very specific roles using a variety of criteria, like Job Title, Role, Industry, Business Size and several other criteria. However, the superpower, if you use Hubspot, is the ability to upload your contacts and companies to the Audience Manager. Once you do that, LinkedIn will then begin to do an 'audience match' among its members in the background. This can take a while; I've seen it take up to two weeks. What is it actually doing?

It is waiting on someone from your audience to log in so it can set a cookie, among other activities. If you look at your company audiences, you will also find that it is identifying every relevant member of any particular company you may have uploaded. It's also capable of creating look-alike audiences with LinkedIn, but you're going to get higher returns with audiences you've attracted organically.

Though the popular refrain is that LinkedIn paid advertising is much more expensive, it is all relative - to your targeting. The ability to be so specific with your targeting, supplemented by on-going content creation and engagement on the platform, will yield huge results for you. That means it is not a platform suitable for a purely paid approach. It means you have to participate as a member of LinkedIn, and back it up with a solid offer and well-designed advertising strategy.

I would go so far as to say that for many businesses, LinkedIn should probably be where they spend all of their advertising dollars. However, that would only be my advice after, and only if, you got your measurement plan in place, and have actually gotten your conversion numbers from running ads. What else does it mean? It means you better have sufficient budget to run paid campaigns before changing your strategy. Nothing is more counterproductive than a client who wants to start making changes before you have this data.

Like all other paid advertising channels, you need to know your numbers in order to achieve success. And like all other ad platforms, this means you have to have enough money to run test until you get results. This means it is not for the faint of heart or shallow pockets. What you're basically always trying to figure out is how many clicks will it take to generate one contact. Then you're trying to figure out how many contacts it will take to generate one sale. This is called a conversion rate, and like all other conversion rates, it can be optimized, but it may or may not be profitable, either in the short or long term.

Thus, as I have said in other chapters and in other books in the series, you must have a working sales process. That means a relentless focus on pursuing every deal, every time, until they either close or qualify out. It also means you have to have a system, typically a CRM, that helps you seamlessly keep track of every contact, throughout the full engagement lifecycle with your company.

Twitter - Use with Caution

Though Twitter saw a huge gain in popularity as a result of its constant use by former President Donald J. Trump, it is a tricky platform to advertise on successfully. Like the previously mentioned platforms, Google, Facebook and LinkedIn, Twitter also has a tracking code that must be set up, which purports to be easy to do. I've done it many times, and can tell that, yes, initially, it is easy to drop in their base code. However, to truly set up a functioning, meaning trackable, digital marketing sales channel on this social media platform, there are quite a few setup steps to actually getting it dialed in properly. I won't provide step-by-step instructions for this one but would be more than glad to assist you in getting it set up.

That said, it can be a very useful channel, especially if you take the time to build a following on it. There are actually at least 12 different ways to engage people on the channel, and it is now no longer limited to just 140 characters. It is now double that, plus it is very easy to use multimedia, i.e., video, documents, audio, on it. All of these activities, for the most part, can be promoted or sponsored. With a system like Hubspot, you can connect up your website to your Twitter Channel and directly tweet content from Hubspot, which provides trackable links for people to click on. This has worked very, very well for me in the past, especially with clients who had very large followings, and who had teams who 'engaged' on the channel on their behalf. In this case, because it was a celebrity with millions of followers, across almost every social media channel, we had dedicated people (in some cases, more than one person) who managed each channel on his behalf.

Twitter, unexpectedly, turned out to work as good as or even better than Facebook in meeting our advertising goals. Speaking of Goals, like all other channels, Twitter works best when you set goals for it, and then spend against those goals. You will almost certainly fail if your goal is 'as much as possible' versus I want to sell X dollars via this channel at this cost per conversion.

Pinterest - Not Just for Moms

Though I do consider it to have much overlap with Facebook's Instagram, Pinterest is a very different 'image' based platform from Instagram. Like before, the first thing you have to put in place before you spend any money on advertising on Pinterest is your tracking codes. Like

Facebook, you also need to design your 'funnel' and model it in your Pinterest Analytics backend, which can be complicated.

Where it absolutely shines is in the B2C consumer space for products that have a visual appeal. This includes fashion, food, real estate, automotive, and a whole host of other industries. To be clear, it's an engagement first platform, and those who simply want to pour money into ads without building a community will experience little success and engender deeply negative consumer responses.

Summary and 10 Lessons Learned

There are many other social media channels to spend your money on, including Twitter, Pinterest and LinkedIn, to name just 3. Like Google PPC, the key to success is knowing your numbers, in particular your sales conversion rate numbers. What you should keep in mind about these channels:

- **Find Your Best Audience.** They allow you to target very different audiences.
- **Use Their Targeting Effectively.** They provide you targeting abilities that are highly granular.
- **Ensure Each Platform's Tracking Code is Installed Correctly.** Each has its own tracking code that must be set up and tweaked.
- **Work For Both B2C and B2B Type Business.** Some are more suitable to one type of business than the others.
- **The B2B Champ.** LinkedIn is primarily known as a B2B channel.
- **You Can Sell on LinkedIn's Service Marketplace.** LinkedIn also has a service marketplace, but it is still very limited in its capabilities.
- **Target by Interest on Facebook.** Facebook has profound depth when it comes to interest targeting abilities.
- **Combine Facebook and Chatbots.** Facebook Messenger is highly effective, if you know how to build a good chatbot that resides within it. By itself, it is still mostly just an instant messenger application.
- **Maybe Twitter Can Work for You.** Twitter can be an effective channel, but it is very challenging for most businesses to effectively use.
- **Each Channel Will Require Sufficient, Skilled Human Resources.** For many if not all businesses, it will take dedicated staffing to leverage these channels. Be careful to not waste time on channels that aren't working.

CHAPTER FIFTEEN

Marketing Spend

HOW MUCH TO SPEND ON MARKETING?

As a digital marketing expert, most of what I am going to tell you is from the perspective of a fully digitally enabled marketing funnel.[57] At this point in time, almost every business either is digital or is in the process of digital transformation (or should be), which can mean a lot of different things to different business owners.

Many Channels - Complex Choices

In days of old, say, when my dad, who was a water well driller as was I, wanted to advertise his small business, his marketing spend was very minimal, and to be honest, very simple. First, he ran an ad in the yellow pages, for nearly 40 years, never spending more than 40 or so dollars a month on it. He had some business cards, and some magnetic signs that got put on his work truck and well rig (which, frankly speaking, didn't last long in the harsh environment we worked in).

His business, after many years, came to depend on word of mouth and that ad. But either because he wouldn't or couldn't grow it, it never got much beyond a one-man operation, even in the best of times. In the worst of times, he had to take another job, as it happens, as a deputy sheriff (today we would call that his side gig). During that period of time, I ran the rig, and was actually a licensed water well driller. I was able to make a go of it, much to his chagrin. I was actually making more than most of my teachers it turned out and more than him.

His business, though, was just like any other business. It depended on getting the phone to ring, and then following up, not an easy thing to make happen consistently. But happen it did, and even though he has passed away now, people still call because they need water. Unfortunately, there's no one left to respond, so we pass the opportunity along to other drillers.

His was a story of both limited customer demand, due to very little

[57] This is simply a way of saying you have a marketing system that is mostly digital, from top-to-bottom. However, for many high-ticket, longer-term sales cycle products and services, this is almost always a blended funnel, with many offline steps that escape measurement, such as custom pre-sales demonstrations of ERP software.

marketing spend and limited capacity, due to the technology he was using to drill wells with, which was based on old 'cable tool rigs', that drilled very slowly.

It should also be noted that regardless of which technology is used to drill, the outcome is the same, a hole in the ground. I used to like to joke that we made nothing from something, because that is what a hole in the ground is.

Capacity versus Demand Planning

Over the years, he had repeatedly pursued upgrading to the latest or at least slightly newer technology, called rotary drills. The difference in cost and performance were dramatic. For instance, his old Stone Model 31 Cable Tool rig could, if the weather held, drill about 150 feet in 3 days in the region (Southern Indiana) where he plied his trade. That rig, bought used for $2,700, is still usable today, and still able to drill at about the same rate. Compare that to the rotary rigs that were and are available which are suitable and capable for the geography of Southern Indiana. At a minimum, they cost about $350,000 out of the gate, without the equipment you need to go with them nor of the accompanying service trucks. All in, if you went new, count on $500,000 to get started. That sounds like a lot of money for a small businessman to pony up, and it is.

However, that new rig could drill those 150 feet in about 4 hours, conservatively speaking, meaning it could drill two 150 foot deep wells every day of the week. Operating only about 10 hours a day including move and set-up times. So, let's do some math, so we can figure out how much he would have to have spent on marketing to make it all work.

On the old cable tool rig, realistically speaking, he was never drilling more than 1 well a week, as you had to tear down and set up on either side of a job. That comes out to be a maximum of 52 wells a year that he could drill. Unfortunately, the weather in Indiana is unsuitable for working outdoors about 6 months out of the year if your equipment is not specially designed and built to withstand it. Realistically, there was only about 20 to 22 weeks a year of drilling time available, which would mean a maximum of 22 wells could be produced by his setup. But it also rains a lot in the summer, it's often hot, and old equipment breaks and not all 150-foot wells can be drilled in 3 days. Then there are other distractions, such as mushroom season and fishing and hunting season(s).

From experience, the most he ever was able to drill was about 20 wells a year. At 10 bucks a foot[58], 150 feet deep and 20 wells, his maximum income was about $30,000 a year from drilling wells. As it happens, this is about what he did, for nearly 40 years. The only time he was able to exceed that revenue figure, was when we had two rigs running or we were able to raise our prices, which was very difficult to do, given the perceived competition, not just from other drillers, but from water lines, ponds, and cisterns.

Now, let's take the rotary, which we were never able to acquire, not because the banks would not finance, but because we could never see the numbers clearly. The rotary came equipped to operate in all weather, year-round, and was highly automated, meaning you didn't need labor other than the operator. It could drill two wells a day, including the setup and tear down, as it was completely hydraulic, and the setup was a 15-minute affair, involving pressing a few buttons and pulling a few levers. Able competitors were all around us, and they did drill 2 wells a day, year-round. So, the demand was and is still there. That's 104 wells a year or about $156,000 a year in potential income. As it happens, the payment on something like that would probably have been around $6000 a month. Your yearly payment then totals $72,000, while gross profit would be $156,000-$72,000, which leaves $84,000 to pay yourself, maintenance expenses and marketing expenses as well as taxes and all the other expenses of running a business, which are many and complicated.

Now we have some simple "rules of thumb" when it comes to estimating required marketing spend. The simplest of all is to use a percentage of revenue. For a slow, conservative business like well drilling, that might not ever go above 10% (or far less, depending on how established it was), while for a fast-growing company, like a SaaS startup or even old-line companies, like Revlon, the number might well approach 30%[59].

Let's keep it focused on the drilling business case study. It's very interesting to me because: 1. I worked in the business and 2. Made my decision early on to go join the Air Force rather than continue on in the

---

[58] That's a 'high' average as for many years, it was quite a bit below that rate, and for a few years, it was considerably above that rate.

[59] Which is how much they declared they spent specifically on marketing in one of their more recent annual reports. Please note, they are an old, very established company and this seems to be an extremely aggressive marketing spend. However, the market they are in is highly competitive and very profitable and virtually recession proof.

family business, which, as you might imagine, was received with mixed feelings, none of which were too positive.

By the Numbers

To keep this exceedingly simple, let's use percentage of revenue to figure out how much to spend. On an annual $156,000 revenue business, admittedly, a very small business, a 10% marketing spend would have come out to $15,000 a year on marketing spend, far more than was required to generate sales for the cable tool rig. However, if you take the $84,000 Gross Revenue and subtract out $15,000 in marketing expenses, as you should, you're still left with $69,000 in Gross Revenue, a little more than double what he was making working only 20 to 22 weeks a year using a piece of 70-year-old equipment.

By the way, why 150 feet? Why not just drill one 300 foot hole in the ground? Because in that part of the world, there is a layer of 'stink water[60]' that if you perforate it, will ruin your well. As you go deeper, you also will usually hit a layer of salt water, rendering the well useless. Just thought you would want to know. There are other areas of the United States where they drill for thousands of feet to get water. But not in the hill country of Indiana. There are also significant quantities of oil deeper down in certain areas, which will also ruin your water well.

Walking Up the Revenue Ladder

Ok, so that's 'small fry' numbers, let's go the other direction and take a look at a business that is in the $1,000,000 to $5,000,000 range, which is still considered small by the United States Small Business Administration or SBA. The exercise is again, one of using 'rules of thumb' and the data you know. Let's use percentage of revenue this time, as we did in the earlier example.

To keep our analysis consistent with our previous UP consulting example, we want to achieve $2,500,000 Gross Revenue and $1,000,000 Net Profit. From our previous calculations, based on an AOV $100,000 and an annual Gross Revenue Number of $2,500,000, we know we need to generate 25 Customers. Based on our paltry .02% Sales Conversion Rate,

---

[60] The water contains heavy concentrations of sulfur.

we also know we need to generate at least 50 leads per customer, or 1,250 leads for the year.

If we use our 10% rule of thumb, that means that in order to generate those 1,250 leads, we are going to need to spend[61] at least $250,000 on lead acquisition. Now we have some top-level data to work with. We can spend no more than $200 per lead. That's actually a very generous number to work with and about 2 times the average cost per lead as reported by Hubspot Research[62]. However, in order to truly reflect the cost per lead acquisition cost, you also need to take into account such things as Hubspot subscriptions, content production, personnel costs, and anything else that goes into your marketing mix. Much of the fixed costs keeps producing leads long after the initial investment, while a paid lead acquisition strategy, aka, renting eyeballs[63], means you will have to continuously invest in your paid channels.

What should jump out at you is that I am using a linear approach. That means I am assuming you will make the same amount on each sale. This is almost never the case. As I've detailed before, I've specialized in the past in selling SAP into Very Large Accounts. However, I was responsible for selling, at least within my industry sectors, into all accounts. That's where a partner strategy comes into play, and when managed properly, can result in very strong revenues.

This is where lead segmentation comes into play. If I am pursuing a deal worth $20,000,000, I am much more willing to spend thousands of dollars to acquire a single lead than if I am pursuing a $100,000 deal. Paraphrasing a famous quote, all deals are equal, but some deals are more equal than others.

Spend As Much as You Can

What all this boils down to is once you know what your Sales Funnel

---

[61] I use 'need to spend' because for many small enterprises, they will try desperately to spend far less and get the same results, usually failing.

[62] Hubspot Research, https://blog.hubspot.com/?_ga=2.160798525.1230005888.1627900344-425174373.1623265371

[63] Google PPC is essentially the art of paying for the same person to see your ad over and over again until they click.

Conversion Rate numbers are, as long as they are profitable, the right answer is to spend as much as you can, for as long as you can. To put it in simple terms, if you have built a system that allows you to put a dollar in and get 2 back out, then the only realistic answer is "how many dollars can you put in". You better do it quick before competitors catch on.

How Would Competitors Catch On?

First off, if you find yourself in an environment that is competitor free, consider yourself lucky. It won't last. If, like almost every business, you find yourself facing a multitude of both direct and indirect competitors, you must know that they have a lot of ways of spying on you. Let's just discuss one, which I myself use all the time (and which I am certified[64] on), SEMRush.

SEMRush or Search Engine Marketing Rush, is a tool that provides many functions for the digital marketer. When it comes to competitor research, it provides a whole host of tools that allow you to spy on competitors. Probably the most important one is that it identifies competitors that you often do not even realize you have.

By using this tool, I can see what words and phrases competitors are targeting. It also allows me to see which competitors are targeting which words and phrases. It also allows me to see the ads they are running for those phrases. That means I can target the same phrases or avoid the same phrases, both strategies are legitimate approaches to beating your competitors.

Pro-Tip - Improve Your Offer for the Win!

It can get very expensive to directly attack a competitor. A much wiser and almost certainly cheaper approach is to improve your offer and your ad. If all you do is copy your competitor's offer and ad-copy, which a lot of 'budget' PPC agencies do, you will find yourself on the losing end of a bidding war. The good thing is SEMRush can help you detect when you're in such a situation. The bad thing is, it is neither commercially viable nor strategically sensible to go this route. So build a better a mousetrap and let the world know!

---

[64] I hold multiple SEMRush certifications, so I guess that makes me a certified competitive spy.

Summary and 10 Lessons Learned

Knowing how much to spend on advertising is critical. In order to arrive at this number, you need to know your conversion rates. Above all, your marketing spend needs to align with your overall revenue targets. What you need to keep in mind when it comes to determining how much to spend on marketing:

- **Depends on Having and Setting a Revenue Target.** Without this target, any spend will be perceived as too much.
- **Match Delivery Capability to Marketing Spend Levels.** Your delivery capabilities also influences, to a degree not often appreciated, your marketing spend.
- **Failure to Plan is Planning to Fail.** Never assume that if you sell it, you will find the spare capacity to deliver it. That's what annual business planning is about.
- **Know What It Cost to Acquire One Customer.** Once you know how much it is costing you to acquire one customer, then you know how much it will cost you to acquire however many customers you need to acquire to hit your revenue goals.
- **Know If You Should Outspend Your Competitors.** If you can spend more than your competitors to acquire a customer, you probably should.
- **Know Your Customer's True Value to Your Business.** For many products, there will only ever be one sale to a customer. If you don't capture that sale, you won't get a second chance for a win-back.
- **Know How Many Competitors You Have?** You have many competitors. You may not know who they all are. With the right tools, you can see who Google thinks are your competitors, which is often not who you think it is.
- **Perform Competitive Analysis on Competitors.** There are many tools available that help you spy on competitors.

- **Assume You Will Be Competitively Analyzed.** You should assume they are spying on you using those exact tools.
- **Having a Better Offer Helps.** That is why, above all, focus on your offer. While the best offer may not always win, if you hope to win, it sure helps.

CHAPTER SIXTEEN

Value of Each Customer

ALL CUSTOMERS ARE NOT CREATED EQUAL

While you should treat every customer with respect, not every customer has the same value to your business. That's why most businesses have a differentiated pricing strategy, usually along the lines of low, medium, and high cost. Within that range, however, there's a universe of different pricing strategies and more importantly, differently valued customers.

One Time Customer versus Repeating Customer

In previous chapters, I've discussed the math behind setting actual revenue goals. I've also helped you understand how many contacts you're going to need to acquire to generate one customer. Finally, I've shown you how you can use some simple 'rules of thumb' to estimate very accurately what it will cost you to acquire the number of customers you are going to need to hit your annual revenue targets. I bet you didn't realize I had walked you through all that fairly complex process, but you've done great. You've kept up with me so far. I hope you'll hang on a little while longer.

Now I am going to tell you one of the most widely mis-understood secrets in the world. That secret is very simple yet critically important to your success. What is that secret?

It's a lot easier to sell to an existing customer than it is to sell to a brand-new customer. It's not just easier, it's far cheaper. There's a reason Amazon was able to build a business based largely on customer reviews. Those reviews are trusted, by a large margin, far more than any information a customer is likely to get from you, a vendor.

Buried within this now widely known insight is the fact it cost about 10 times[65] as much to acquire a new customer as it does to upsell or cross-sell into an existing customer. This is true whether you're talking small dollar purchases or Very High-Ticket items, like SAP.

---

[65] This is actual data I collected while selling SAP. We had highly granular Business Intelligence data available. It was very clear, it easily cost 10 times more to acquire and close a new customer than it did to sell something to an existing customer. It cost more money and much more time.

The Leverage Recurring Subscriptions Bring

Before I get down to the math (you knew there would be math, there always is when it comes to marketing and sales), let me explain what a recurring subscription is, and give you a flavor of just how many varieties of this type of sale there is.

At the most basic level, a subscription is something like a newspaper subscription. It can be physical or digital. But the most important aspect of the model is the fact that once you sign-up for a subscription, you will usually keep subscribing for years and years. On the next higher level are subscriptions that don't appear to be subscriptions. Electricity, Phone, Water, Internet. All of the 'essentials' of life. You sign-up and consume it constantly. The unique thing about these business models is that when they raise prices, they encourage you to use less of their service! They never tell you they were successful at lowering their cost of production and are therefore glad to pass along the savings. Except for the phone company, who sort of took a major hit when the internet caused the cost of long-distance phone calls to collapse.

But these 'essentials', such as electricity, just keep going up and up. Their utility doesn't increase, but their price does. That's why your grandma was advised to invest in utilities. At least until Enron messed up the show a little bit.

But there are a whole host of consumer subscriptions these days, primarily characterized by being facilitated by ecommerce platforms, like Shopify or Amazon. That's why just about every business I've worked with is always trying to figure out their subscription model. It's just a lot cheaper to sell to someone who is consuming via a subscription model.

But let's move back up to where the big boys play, namely, SAP. You see, SAP makes a lot of money up front and on the backend. There are two main back-end channels to be aware of - especially when thinking about your own model. The first is support, currently 22% of the license cost (list, not discounted, a major point to keep in mind). At that rate, you buy the license again every 4.5 five years. It's actually a good deal for customers, at least in my opinion, though your mileage may vary. The other major 'subscription' component of the SAP model is selling additional seats or solutions into an existing account, otherwise known as filling in the whitespace.

You might ask, why didn't the customer buy everything they needed to start with? As usual, cost is the culprit. Up until not that long ago, a typical SAP sale would consist of no less than 50 user seats, list price about

$300,000, maybe half of that after discounts. That would get you coverage for your core Financials, Human Resources and Logistics processes[66]. That's already a lot to chew on for most companies. However, for many companies, if not most, shortly after they go live, they immediately are faced with increasing the SAP footprint to meet new requirements, whether parked[67] during the first project for various reasons or discovered during implementation and deferred.

LTV:CAC Magic Numbers

As I mentioned, there's always math involved. Though I've walked you through various subscription models that range from the cheapest to the most expensive scenarios, i.e., Newspaper Subscriptions and massive multi-million-dollar ERP projects, the basic math that makes the subscription model (by whatever name a company prefers to call it) is straight forward. What you're looking to find out is the ratio between the Lifetime Value of a Customer (LTV) and the Cost of Customer Acquisition (CAC).

You're looking for not just any ratio, but rather, a ratio of at least 3 to 1. In other words, you're looking for the Lifetime Value of a Customer to be at least 3 times greater than the cost you expended to acquire that customer. That's just a rule of thumb, and you may well need to achieve a higher ratio, or you might be able to survive at a lower ratio. However, you should know that Hubspot, our Inbound Marketing and Sales software supplier, a SaaS vendor, nearly went out of business because they initially didn't get this number right.

They had a product that was priced very low. It was new to the market, with very few competitors. It also had very limited functionality. But, in our case, it had more than we needed to get started as we were not Inbound Marketing experts. Now we are.

Sometime around 2012, they hit the wall. Their LTV:CAC ratio was about

---

[66] SAP's ERP solution, at least the core modules, consist of 16 modules. But they have many other solutions you can buy, and their partners have many hundreds of others you might end up buying. It all comes down to what is needed to accomplish your goals. But just about everybody starts with the three mentioned core modules (but not 100%).

[67] As a SAP Project Manager, you will be faced with the difficult choice of saying no to a requirements change now, but will park it for later incorporation into the solution footprint, when there is more budget and the change can be justified.

1.5 and they were bleeding money. At the same time, they were investing heavily in R&D. From the time we subscribed, at about $200 a month, up to that point in time, they had released two new products, a professional version and an enterprise version. But their LTV:CAC ratio was not sustainable. As a result, they not only released these new differentiated products, but they also raised prices. They also shifted their sales effort toward bringing in more enterprise class customers.

Why? Simple, the small enterprises, the one- and two-man bands, simply don't have the revenue nor personnel to fully leverage the Inbound Marketing methodology. Here's what Hubspot's numbers looked like before and after the shift.

This model is not specific to Hubspot, it just happens that I have their numbers from a publicly available presentation

Hubspot LTV:CAC	*Hubspot in 2012 LTV CAC = 1.6 (Not Viable)*	*Hubspot in 2015 LTV CAC = 2.43 (Viable)*
CAC (Customer Acquisition Cost)	*$6,025*	*$6,880*
MRR Per Customer (Monthly Recurring Revenue)	$429	$583
Churn Rate	3.5%	2%
Margin	83%	81%
LTV (Lifetime Value)[68]	$10,173	$23,612
LTV:CAC[69]	1.69	3.43

Table 43. Hubspot Initial vs.. Final LTV:CAC Values

---

[68] The LTV is determined from the following formula. MRR/(Churn Rate*Margin)

[69] The LTV:CAC ratio is LTV/CAC

This is a very powerful formula, not just for SaaS companies. If you have a business model that includes MRR (Monthly Recurring Revenue), this tells you whether your model is financially viable. If your business does not generate any recurring revenue, it should spur you to consider ways of turning it into an MRR model.

There has been an explosion of businesses developing or redesigning their business around this very concept. There are several challenges to overcome when transforming your business model into a MRR model. One of the key ones is your IT system(s). I'll cover some of the requirements your IT landscape is going to need to have in a later chapter. The other is your Human Resources approach. It's very different to build a sales team that hunts versus one that farms. I know, I know, that's a 'dated' view of sales, but it is still very valid. I am going to cover some of the HR challenges in a later chapter as well.

One the biggest challenges with transitioning to an MRR model is your marketing approach. It's one thing to attract a brand-new customer, it's a completely different challenge to keep them sold and to continuously sell them new products. Your digital marketing landscape has to be ready to handle all aspects of this approach.

Because the digital marketing landscape has absolutely exploded with new systems, some 8,000 and counting since 2010, the skills needed by your marketers as well as your sales team have dramatically changed. I am going to delve into some successful approaches I've employed over the years to find and train new marketers and salespeople. Be advised, universities are way behind the power curve when it comes to teaching these new approaches.

Perhaps the biggest challenge of all is what we colloquially call the "No Budget" problem. As you might imagine, all of these new digital marketing systems and strategies come at a price, a very steep price. No longer will a small ad in the yellow pages suffice for most businesses. In fact, you would be hard pressed to even find a physical copy of an up-to-date copy of the yellow pages phone book in most houses. By reading this book, and others in the series, I hope to bring all these formulas, rules of thumb and strategies together to help you put together a realistic marketing budget that helps you meet your Growth and Revenue goals.

Finally, I want to also warn you about the multi-system patchwork mess of systems problems. If you have not guessed, all of these differing systems introduce a lot of complexity into your environment. Even a small business may have 10 subscription SaaS products all running at the same time. Larger businesses will routinely have 50 or more. This induces a headache of epic proportions into your business.

Summary and 10 Lessons Learned

Every customer is important. But not every customer is equal. The cost of marketing and related IT systems has soared, and you should expect this trend to continue. Digital marketing doesn't come cheap. But when you get your model right, profits can soar. What you must get right:

- **Implement a Subscription Model.** A large number of businesses have successfully found ways to convert their business model into a subscription model.
- **Know Your CAC.** You must know the Cost of Customer Acquisition. This is complicated to get right but can be done, even by the smallest operator.
- **Know Your LTV.** You must know what the Lifetime Value of a Customer is.
- **LTV:CAC Ratio is Key.** You are looking for the LTV:CAC Ratio to be at least 3 or above, generally speaking.
- **Attack Churn.** Treat your customers right and reduce churn.
- **Reduce Churn and Achieve Leveraged Results.** A small difference in your customer churn rate can have an outsized impact on your LTV.
- **Make Use of Upsells and Cross-Sells.** Most successful subscription models also offer Upsell and Cross-Sell opportunities.
- **IT Systems Must Be Capable of Supporting a Subscription Model.** Your ability to pull off a subscription model successfully will frequently be limited by your IT systems.
- **Budget Appropriately.** You must have sufficient budget to pull off this model.
- **Mine Your Existing Customer Base.** An existing customer is much, much cheaper to sell something to than to a completely new customer. As much as 10X cheaper! The timeline to make a sale to an existing customer is almost always far shorter than it is to make a sale to brand new customer.

CHAPTER SEVENTEEN

Selling Your Plan

KNOW YOUR NUMBERS TO SELL YOUR PLAN

So far, I've walked you through a step-by-step process for determining how many leads you're going to need marketing to generate. I've shown you how that number is derived from having a Top-Level Revenue Goal in mind and then working backwards through Average Order Value (AOV).

The Red Thread Connecting It All

Running throughout this series of calculations is a common thread, some call it the 'Red Thread'. That is that at every stage of the process, there's a conversion rate involved. You can run your numbers from the Top-to-Bottom and you can run them in reverse, from the Bottom-to-Top. One will tell you how many of everything, Contacts, Leads and ultimately, Sales you need to hit your revenue target.

The other one, the bottom up one, will tell you how many of everything you 'must' achieve based on your current, actual results and it can be used to tell you how many of each you much achieve to hit your revenue target! Therein lies the challenge, finding those current results. It's the 'key' reason you need a system like Hubspot. Google Analytics won't do it. It will do a lot of useful things from an analytics stand-point. But what it won't do is tell you which specific customer came from which specific ad, keyword, event, or other activity. In fact, it does just the opposite, and continues to do just the opposite in ever increasing doses due to a focus on data privacy.

Which brings me to why you need a marketing platform like Hubspot. You see, it is, above all else, a measurement platform. With it, you can truly tell, at a glance, what your Lead-to-Sale conversion rate is, by source, and who specifically converted on what. This is hugely important. There's another platform that is even more powerful for this purpose. Funnelytics, which allows you to track every funnel step visually and run simulations. Unfortunately, it starts at about $1,000 a month but it is worth it if your LTV:CAC numbers are aligned[70]. Otherwise, you're going to spawn spreadsheets like Dandelions spawning in the Spring.

That's why I recommend you establish your SMART Goals[71], meaning

---

[70] By aligned, I mean has a ratio of at least 3 or above.

[71] SMART is an acronym for Specific, Measurable, Achievable, Realistic and

set your top line revenue goal, then work your way all the way down to the specific marketing and sales activities you will need to do during the period of time encompassed by your SMART goals, usually annually.

Importance of Goals

Using our fictitious UP Consulting company, that means we break it down from our top-level revenue goal, all the way down to a fully funded content plan, advertising budget, sales incentive plan, and everything in between. This is a non-trivial planning exercise. But I can assure you, companies like SAP invest heavily in this exercise, at every level of the business. I and my extended team routinely spent six weeks performing this planning exercise. Even then, many elements of my plan might not get funded, which meant, in turn, I had to go back and alter my revenue assumptions. As you might imagine, telling your boss that you're unable to commit to the target revenue he set because marketing won't fund some content you planned on having is a tough environment. But that's why you do the planning exercise.

Your plan, once built, polished, and delivered, is just the start of the process of 'selling your plan'. You still have to get business buy-in and most importantly, funding. You see, as part of your planning process, not only do you need to come up with those Contacts, Lead and Buyer numbers and conversion rates, you also have to do some work to put a price to your plan. When it comes to marketing and sales, that means researching the probable cost of any systems you want to procure or subscribe to, as well as any upgrades you will want for existing systems, such as Hubspot or Salesforce. Then there is content production. Do not deceive yourself into thinking you will be able to produce it all by yourself or even with only in-house resources. If you also plan to 'have a booth' at a marketing event, you will need to include those numbers in your budget. There are many, many places where a good marketing plan can target budget.

In my own case, I've created well over 400 pieces of highly specific, targeted content to support my business[72]' marketing goals. But in order to do that, I first had to have decades of accumulated knowledge and experience within my area of expertise. Even then, it took me many years to

---

Timebound

[72] My business being SAP BW Consulting, Inc.

produce it all. As a validation exercise, I and my team performed an exercise where we got prices for every single piece of content we produced ourselves using commercial content production houses who provide publicly available pricing.

The result: It would have cost us a little over $947,000.00 to have it done. That would have been well outside of our available budget at the time, however, we would have earned it all back ten times over - but a lot faster.

That's why I always tell my customers that all of the successful customers I've had started with a content plan. Those that did not develop one (whether with us or on their own), usually failed to achieve the desired results. Almost invariably, they ultimately would say marketing wasn't working. On the other hand, it is entirely possible to not only develop a content plan, but it is also possible to calculate the number of pieces of content you're going to need and to know what the expected return on that content will be. We've even built a content planning calculator to support this calculation.

By the way, there's an overwhelming amount of content out there, and millions of pieces of it being produced every day. The secret to success is that your content has to support what you sell, at an extremely high-quality level. In other words, yours must be better than your competitors.

Of course, the larger the organization, the more complex the organization will be. From a planning standpoint, that means there will probably be multiple planning levels within the organization, and there will be a challenge you must overcome. That's the challenge of rolling up those plans. Within smaller organizations, that's going to mean excel, unfortunately. Within larger organizations, at least those that use SAP, they will most likely be using either the SAP BPC[73] or SAP SEM[74] modules to complete this planning exercise, at least the financial part of it.

As it happens, I am a SAP Certified Strategic Enterprise Management (SEM) Consultant[75]. I therefore lean on that background to discuss the

---

[73] SAP Business Planning and Consolidation. SAP BPC is a Business Planning module available from SAP

[74] SAP Strategic Enterprise Management

[75] By certified, it means I attended a 5-week SAP Academy on the subject and took and passed the Certification Test.

planning process and selling your plan. In truth, although these modules are stand-alone, they are deeply integrated with various SAP modules. I will mention these here, but you should realize that you probably have the same basic functionality within your organization or should have.

With SAP SEM/BPC, each business unit performs planning all the way down to the individual product level. This is not a problem when you have only a few products, as is the case for many small businesses. But once you start having hundreds or thousands of different products and services, being sold, delivered and serviced on a global basis, you have a huge planning challenge. These tools allow each business unit manager to plan sales at a daily, monthly, quarterly and annual granularity level. These plans can then be automatically rolled up to each geographically distinct or sales planning channel level. It really comes down to whatever makes sense for the business (and this can change quickly over time).

Why Bother with All This Planning?

There are a variety of critical outcomes that come from this planning exercise. Probably the most important of which is predictability. Most small businesses are stuck in the sell 'as much as possible' or anything mode. There's not a lot of resources to plan with it, and most decisions are made by the owner or small group of founders.

But once you have grown into a huge multi-national enterprise, this is no longer a feasible approach to planning. Thus, everybody prepares and submits annual business plans, for their little slice of the business. That means, for example, the four or five revenue-generating elements of a particular region, say Dubai or Spain, each prepare their own business plan and hit submit. These individual plans would then get rolled up to the country level, which then get rolled up to the regional level, and ultimately, to the global level[76].

This is no small challenge, and these processes typically take years to get fully implemented and stabilized. However, I described the process from a bottom-up perspective. The reality is, there is almost always a

[76] This is only one possible roll-up scenario. Other businesses may well be organized along other lines, such as by line of business or product category, or even profit versus cost centers. Sometimes, they will make plans using different organizational approaches to see what the differences are.

'growth' element imposed from top down. If corporate says to the public we are going to grow top-line revenue by 10% in the upcoming year, that number will be rolled down across all planning elements, though it may well not be evenly distributed[77].

Sales Goals Tend to Drive it All

In my experience, most companies tend to have some sort of internal sales hierarchy, within which, Sales Goals are set, and from this, all other planned metrics and related expenditures flow. I will, as always, use my own experience to walk you through how this works and hopefully, provide you with some ideas for modeling your own planning process.

In my role as an SAP Industry Principal, a quota carrying sales role, I had an annual sales quota for net new license sales. I actually 'owned' 4 primary industries, plus 2 more 'orphans[78]'. My annual target was set at $4,500,000 in net new license sales[79]. My bonus was tied to achieving this number. In other words, the company was holding me accountable for generating $2,163 dollars in revenue[80] per hour! My own internal goal was to achieve at a minimum of 5 times my OTE[81] bonus number, which breaks out to $22,500,000 in net new license revenue. In other words, the company was counting on me to bring in $10,817 an hour to the company. Of course, I got paid more than just my bonus, but this was a handy set of metrics to work with for my planning exercise and my eventual 'selling of my plan' exercise.

---

[77] What you typically see here is a tendency to milk the cash cows while spending very little on marketing and sales to do it. At the same time, for the up and comers within a business, you may see large increases in marketing and spend to achieve various objectives, such as market share growth.

[78] Orphan in this case means the company was unable to hire someone with industry expertise to take on the role. As I have cross-industry expertise, then I was able to cover these, at least within my assigned territories.

[79] Net New License Sales means to a brand-new customer who has never previously purchased a product from SAP. These are both the hardest sales to make as well as the most expensive. However, they are the most critical sales to make for long term growth.

[80] Assumes 2080 hours of available work a year. Does not account for utilization rates as that does not apply here.

[81] OTE - On Target Earnings, that sales figure at which one earns 100% of the agreed upon annual bonus.

No One Rides Alone

I want to be clear here, I didn't work alone. Instead, I worked within a matrixed[82] organization structure, and had many people providing me a small portion of their services, most importantly, there was a small army of Account Executives who also carried a quota and who depended upon me to make sure their and my pipeline was full. But full was precisely defined. As yours should be. What did full mean? You will run across many different terms for this, but what it was known as within my world in this environment was pipeline coverage.

Pipeline coverage means how much opportunity you have in your sales pipeline, in dollar terms. When I first started out in the business, we knew we needed to have at least 3X pipeline coverage, but that has now increased to having 4X pipeline coverage. What does that translate into English as? 3X pipeline coverage means that if you have an annual sales quota of $4,500,000, you need to have $13,500,000 in your pipe. When it increased to needing 4X, it meant I now needed $18,000,000 in pipeline coverage.

Remember, I had a personal target (but one which was, in fact, pushed upon me by my manager) of achieving 500% of my bonus. Let's work those numbers so we have context. We'll use 4X pipeline coverage here. What does that now change my number to? $90,000,000, which is 5 * $18,000,000. This then, was how much "opportunity" I needed to have in my pipeline, at any one time. However, this is actually a rolling average. Pipeline opportunities are continuously getting 'consumed' by either being won or lost.

Probability Weighted Sales Pipeline

As I 've previously explained, each stage of your pipeline has a conversion rate from the previous stage. As it turns out, most CRM software packages allow you to create your sales pipeline within the system and assign a probability of closing the deal to each stage of the pipeline. Now, admittedly, this can be a contentious exercise when setting this up. Small companies tend to do a lot of guessing as to what their sales pipeline

---

[82] Matrixed is one of several common organizational structures. It just means you have many bosses, unfortunately.

actually consist of, and have little or no information[83] about what the probability of closing each stage represents.

Large companies tend to have a lot of information, but also a lot of noise in their pipeline. Though almost every sales trainer[84], including myself, will be pushing you to simplify your pipeline as much as possible, complex sales processes inevitably lead to overly complex sales pipelines.

Your Sales Plan Gets Sold Many Times

Whether you are using a sophisticated system like SAP BPC, or a simple Excel spreadsheet, you need to be able to sell your plan. Even if only to yourself. How does yours need to be sold? Communicate, communicate, communicate.

I've already explained the 'must have' numbers in a plan:

1. Top line revenue goal
2. Leads Required
3. Conversion Rates.

When I've presented my annual plans, the process was always the same, present my numbers, then let my boss try to tear them apart. If and when we all agreed on the first three items (which is a lot easier to do if you have an IT system that reveals it), then it was time for the 'ask'.

What's the 'ask'? The money you will need to fund the activities you plan on undertaking throughout the year to support your plan. This is usually where the battles are fought. But if you work in an organization or are trying to build one, that is working from agreed upon top level business objectives, with agreed upon KPIs (Key Performance Indicators), and measurement systems, you can turn your pitch into a collaborative, less friction filled affair.

---

[83] However, once you start to make sales, the system will tell you what your actual conversion rates are between stages and you can use this information to update your deal stage probabilities.

[84] I don't formally call myself a Sales Trainer nor do I market my Sales Trainer services. However, I have trained a large number of Account Executives over the years and have a very good track record of creating people who go to the President's Club or its equivalent.

What then, should form the outline of your plan:

- Annual Revenue Goal
- Average Order Value
- What you plan to sell, usually broken down by product or service line
- Who your top 10 target accounts are
- Planned marketing activities with cost estimates
- Planned Sales Activities
- Delivery Plan (can be a verification that production can meet the demand or that you can deliver the service or both)
- Finally, ask for approval.

That's your basic pitch deck. Once this is approved, the next step is to set the entire plan up in your CRM system. Sticking with my assumption that you have a system, my recommendation is that you take some time to plan out your activity account-by-account. That means you and your sales team are going to need to invest significant time planning out specific activities for each account. The larger the account, the more time you will need to invest in this planning process.

Some people call this TAS© or Target Account Selling[85]. Whatever you call it, it means you have to plan out, account-by-account, what you will do to get the account over the finish line and turn them into a customer. Many systems, such as SAP, will have a specific tool for this process. I believe SAP's was called the Territory Planning Navigator or TPN[86].

What's special about it? In my experience, it's the other half of selling your plan. You see, if the first pass of selling your plan was up-the-chain, then the second is down the chain, specifically, to the Account Executives. They are the shock troops of your sales process and unless and until they sign up to your sales plan, then it will never happen.

---

[85] I was first introduced to this concept working for i2 Technologies. However, it is currently primarily called that by Siebel, who are now owned by Oracle. You may come to know it as Account Based Marketing or ABM.

[86] This was an internal SAP extension to the CRM tool from SAP.

I found that if I sat down with each one of them individually and worked on specific account plans, whereby we identified each person in the account and each planned activity, then they 'owned' the plan. In other words, we had buy-in.

Though every business and sales planning process is undoubtably different, I found that the Sales Planning Process, whether it was for SAP, i2, Oracle, or my own SAP BW Consulting services was limited to about 8 to 10 accounts a day. That means that if I wanted every Account Executive to have 50 well planned accounts, we were doomed to spending at least 5 days together developing these plans and making sure they were in the account planning tool. It also meant that I had to budget at least two solid months of my time for this exercise to make sure every Account Executive who worked for me had his share of the planning done.

I would like to say it was unnecessary and was just me over planning. Au contraire. The more time I invested in this process, the more likely I was to hit 500%. Almost without exception, if I didn't spend this extremely expensive time planning out the sales activities with each individual Account Executive, neither they nor I hit quota. And I liked hitting my personal quota target.

Keep in mind, I mentioned that your "pipe" was continuously consumed by either wins or losses. That should tell you that this should be an on-going sales management activity throughout the year if you want to hit your numbers. There's one other reason you will need to plan on continually investing this level of effort - staff turnover.

You see, Sales is a high turn-over environment. The really good ones will stick around for a long time, however, the better they are, the more likely they are to get poached by a competitor. The ones who are not such high achievers will tend not to make quota, and will fall victim to the annual 'culling of the dogs.'

This also is one of the reasons you don't generally see a huge amount of product training being given to the AEs. At most, a new guy would get about a week's worth of training, though I believe they have now once again expanded it to a 2 or 3 weeklong sales academy. Realistically, with more than 1300 items on the price list, an AE isn't getting paid to be a product expert. There's an army of Pre-Sales people, value engineers, product marketing specialists and executives who he can bring to bear on the sale.

The challenge is for the AE to know which resources he has available and how to call them in. That's where somebody like an Industry Principal comes in - they are supposed to know the product inside out and have an extensive network. In my experience in this role and other similar roles, the real secret to success was having a vast network of experts who you could tap, including among partners and even customers, to win the deal.

Include Your Partners in Your Plan

For many businesses, SAP being no exception, they are partner dependent. What does that mean with regards to selling your sales plan? Very simple, you're going to need resources from them, either ones they provide 'voluntarily', meaning free, or ones you buy in. In either case, if you want to have who and what you need where, when, and how you need it, you need to spend time planning out accounts with your partners.

Partners are usually asked to commit to a certain amount of revenue, in other words, commit to bringing in a certain amount of revenue, which you can include in your overall plan. But they also frequently come to be the only way to effectively demonstrate or deliver certain specialized functionality. This doesn't just apply to selling complex software either.

I've seen this model across virtually every industry sector I've worked with. A good partner allows you to effectively multiply your size. For instance, one of our partners was actually 'handed off' the development roadmap for one of our key, most complex pieces of software, which is MRO or Maintenance, Repair and Overhaul. This was a huge commercial opportunity for this partner. But it was an even better opportunity for SAP, because now, we had dedicated resources for developing this complex, yet not a huge seller.

Without this approach, given that we were no longer doing much internal development on this product, we were basically dead in the water sales wise. There was no end to sales opportunities, but there is a limited amount of internal development resources within any company, including even mighty SAP.

I mention this only to demonstrate that it is often times necessary to think well outside the box when it comes to developing your sales plan.

Summary and 10 Lessons Learned

No matter what industry you're in, you're going to need to develop and sell your sales plan, even if it is only to yourself in your one-man band size organization. The Sales Planning process can be resource intense and should be driven by top-level revenue goals. In many scenarios, you need partners to succeed. What you should keep in mind when it comes to selling your plan:

- **Revenue Goal is Key.** Key means knowing what your goal is, in both revenue from new customers and existing customers.
- **Knowing Your Deal Stages and Conversion Rates.** You'll either need to know your conversion rates or estimate them. If you're just starting out, you'll need to use informed research to make some educated guesses.
- **Plan in the Correct Sequence.** The planning process should normally flow Top-to-Bottom then Bottom-to-Top and then iterate.
- **Understand Your Sales Planning Maturity.** Mature organizations usually have highly structured Sales Planning processes, while SMBs and startups will have to rely on spreadsheets.
- **Practice Accountability at Every Level.** Every Sales Person is responsible for a calculable amount of revenue. They should know it and so should management.
- **Spend the Time to Do Detailed Level Planning.** Planning down to the individual account level is usually required to sell your plan.
- **Planning Should Drive Budgeting.** Detailed, account level planning also drives marketing budget decisions.
- **Understand Your Pipeline Coverage Requirements.** You need to have enough pipeline coverage in your opportunity pipeline to achieve your sales goals.
- **Lead Production Rate vs. Pipeline Consumption.** Your lead production rate must at least match your pipeline consumption rate.

- **Pipeline Replenishment Rate Must Match Consumption Rate.** Pipeline is constantly consumed and thus must be continuously replenished.

A SYSTEM OF SYSTEMS APPROACH

SECTION III

In the remaining chapters of this book, I am going to do a fairly deep dive into each of the problem areas I identified in my original book in this series, "How to Dominate Any Market: Turbocharging Your Digital Marketing and Sales" and how each of them relates to solving your growth and revenue problem.

Why talk about systems when I've been focused on Sales Planning, Growth and Revenue? Because in order for you to achieve your goals, you're going to need systems, no matter what size your enterprise. Let's dig into each one.

CHAPTER EIGHTEEN

The HR Problem

ROBOTS DON'T DELIVER THE DEAL - YET

You may think a giant company like Amazon or Jet[87] are robotic driven powerhouses, but you would be wrong. They are, in fact, massive organizations with hundreds of thousands of people working at every level of the organization. You, as a consumer, may be only too happy to log-on to your Amazon Prime account, search, click and buy, and be amazed that your product shows up at your front door, often within an hour. What you may not appreciate is the fact that your entire experience with Amazon is a result of near continuous user testing. And how do those tests come to pass? People dream them up. They rely on the data that comes flowing out of the system, but ultimately, they dream them up. They are able to dream up so many different tests that on any given day, it is reported that there may be well over 1,000 tests going on throughout the pages you look at on Amazon.com.

You're probably not an Amazon class company, but you might well want to be or have plans to compete head on with them. Here's the thing. Amazon is a hellish place to get a job. I don't mean it is hell to work there. I mean they put new hires through a hellish recruitment process, all in the name of getting the very best people in the door. And they apparently ruthlessly weed them out if they can't keep up. Though you may or may not be a fan of Amazon, you can't argue with their results. That's the odd thing about the American labor market - they reject a lot of people for a lot of different reasons.

When it comes to sales though, what do I see among the smaller enterprises? They typically suffer from either low-quality resources or not enough resources or both. Worse, I see, as I also see on the marketing side of the house, interns being given jobs which are critical to the company, yet with little to no training. I think it is great to give an intern a work opportunity. However, when it comes to making your growth and revenue goals, I think it is beyond risky. It is catastrophically dumb.

Work On Your Offer

Since I've walked you through how to set revenue targets, then you know I believe the Human Resources problem has a least some element of

[87] Now owned and operated by Wal-Mart

conversion math involved. Perry Marshall[88], of 80/20 fame, provided the best advice I've run across when it comes to solving your sales problem. Work on your offer. That's right, the better your offer, the easier it is for your least skilled salesperson to succeed. It also means that your most skilled salesperson can be much more productive.

How do you work on your offer? The first thing you have to do is thoroughly understand what you sell. For more on that subject, see my book, "Knowing What You Sell, The Key to Dominate Your Market"[89] but let me provide you with the salient points to this sometimes surprisingly difficult task.

There's actually not that much written about this task. But one of the latest ideas to come out is the idea of Jobs to be Done or JtbD[90] for short. JtbB is really just a way of saying, you must ask your customer what job they were trying to accomplish when they hired your product or service. Let me give you a practical example referring back to my well drilling business scenario.

It actually took us a long time to figure out that what we actually did was make nothing from something. You see, a water well is just a hole in the ground. There used to be dirt and rock where the hole is, but after we got done drilling, there was just a hole. It was filled with air and water if you were lucky. If you were not, it was just filled with air and darkness. The people who hired us hoped we would find water, so they could have clean, fresh, plentiful water come out of their taps when they wanted water. There's a lot of different problems people are actually trying to solve when it comes down to this most basic of functions. They need water to wash clothes with, bathe with and to drink. They need it to cook with. Sometimes, they need to water their lawn. In short, water is life. Without it, we all die. But with it, our houses are more valuable, and our lives are easier.

We could have sold a lot of different 'add-ons' to our customers, such as pump installation, filtration systems and annual water well cleaning

[88] Marshall, Perry, 80/20 Sales and Marketing: The Definitive Guide to Working Less and Making More, Entrepreneur Press; Illustrated edition (July 22, 2013)

[89] Ayers, Lonnie, Knowing What You Sell: The Key to Dominate Your Market, May 4th 2021, Amazon ASIN: B0946YYDK2

[90] Clayton Christensen: https://hbswk.hbs.edu/item/clay-christensen-the-theory-of-jobs-to-be-done

services. We actually ended up doing all of those things over the years, but it was always a one off, secondary sale for the main business. Which was, to be clear, making nothing from something.

So, our offer was basically very simple - but not so easy to sell. You see, customers bore some risk when they had us drill them a well. There was the obvious risk (which many had not thought about before calling us) of coming up empty or having a dry hole. There were other less well known risks, which we did our best to advise clients of, for instance, the layer of sulfur water (stink water) that you would typically run into if you drilled beyond a certain depth. Then there was the risk of drilling into a saltwater layer, which would allow saltwater intrusion into your well and ruin the well. There were other risks as well, such as having a perfectly good well go dry because of earthquakes (happened to ours), and having other water quality issues arise. In short, the product has a lot of negatives.

It also had one killer Unique Selling Proposition or USP. It was often your only option depending on where your property was located. The areas where we typically drilled were very remote, with few if any public water lines available to connect up to. A good well would save you a substantial monthly water bill and make your property worth a lot more money. It was also substantially cheaper than what it actually cost to hook up to city water, which people are surprised to learn.

Clearly Articulated Value Proposition Wins Every Time

Want to know why Obamacare has been such a dud? Did you even realize it was such a dud? As it happens, one of my marketing customers makes a good living off of selling Obamacare Plans. But wait, you say! How could that be, don't you just go to the government website and sign-up?

Though you can do it this way, many do not. In fact, tens of thousands of people do not, instead, they hire my client to do it for them, and he gladly does it for free. But he does get paid by the insurance companies whose plans he recommends to people. You see, he sells 'quotes'. Why does his business exist and hundreds of others just like it as well? Because the United States Government isn't in the business of selling insurance. Its website has gotten much better over the years, there is no doubt about that. But I routinely go onto it myself and am simply astounded at how hard it is to actually figure out what to buy.

You see, at the end of the day, they are simply acting as a sales portal for whatever the insurance companies are selling. There are many, many choices to make and making the wrong one can and will bankrupt you. In short, it's a product, insurance, that seems almost purposefully designed to be complicated to buy.

Government Mandated Salesman HR Problem

Now that you know what I consider one of the world's worst products is, but which I do manage to make a bit of revenue from, let me tell you about the Human Resources problem this client faces and how he manages it. You see, to even speak to someone on the phone about health insurance, at least in the state where this client is, you have to have a license from the state for this particular kind of insurance. And it doesn't come cheap nor is it easy to get.

To solve this problem, he has to invest in training, and pay for each person's license fees. This gets very expensive very fast, and it's a problem that has to be resolved every year as there is a huge amount of turn-over among the salespeople, who all, ultimately, work in a call center.

It turns out, though, that all of these numbers are actually fairly predictable. He knows about what his pass/fail rate will be among the trainees each year. He also knows what his sales closure rate is probably going to be. And now, he has (but did not initially) an annual sales target, which breaks down into something like 2000 new customers a month.

No Robots Make These Sales

This is a highly digitized marketing and sales process. Yet not a single sale is made by a robot. I've actually worked with robots in industrial settings many times. They're pretty dumb. They are inflexible. But in certain situations, you can basically pump out profit due to their extreme capabilities. But I've never seen one make a sale. I've seen them assist in making sales, many times. For very complex products, like SAP ERP, Oracle, Microsoft or Hubspot, who would love to sell more at the click of a button, most sales are still very, very labor intense. It requires skilled people, not just skilled salespeople, to make these deals.

And that still utter reliance on people in every aspect of business is what the HR Problem is all about. Yes, good systems design and implementation can make every person more productive. Sometimes, a good system can replace some people, which happens often. For example, did you know that long ago, here in Spain, where this author currently lives, women basically could not work because they had to be in the house when the meter reader came around. That's right, meters were read by humans, and in order for them to do that, they had to get into your house.

Today, smart meters are remotely connected, and you just get a bill. Thousands of meter readers lost their job long ago due to this one innovation. There are countless other examples of process improvements replacing thousands of jobs which we are no longer even aware once existed. Yet, employment keeps rising across the board. It is actually very tough to find good people, no matter which role you're trying to fill.

That's one of the primary reasons many businesses rely on cheap overseas labor or interns. Because they are unwilling or unable to pay top dollar. Mostly, they are unwilling and focus on the wrong metrics.

Measure It Right and Get It Right

That's why I firmly believe and advise my Sales Coaching clients that properly trained and skilled salespeople are absolutely critical to their success. Of course, top salespeople are hard to find and retain. But you also need to have and set firm targets which they are completely onboard with. Otherwise, you will have a very rough time managing your sales force and by extension, hitting your numbers.

Fortunately, the CRM systems available today, even including the free versions such as the one available from Hubspot, allow you to design, monitor, forecast and measure your sales results easily. Thus, that's why it is necessary to set goals. Without it, all sales conversations come down to sell more, more, more versus, you did a great job.

Tie Top Line Revenue Goals to Individual Targets

If you've followed along with the process I've laid out so far for setting growth and revenue goals, then you are ready to set to individual revenue goals across the organization. Notice I said across the organization, not just for your sales organization.

You see, though it is entirely normal to have and set sales quotas (though this is rarely done well, in reality), it is also desirable and necessary to set revenue goals across departments. Everyone needs to know how their activities contribute to the overall goal of the company. Let me give you a practical example, from a world I have more experience in that most - delivering professional services.

I've already walked you through the process of working backwards from a top line revenue goal all the way to the number of customers required to hit that goal as well as how to set up and measure a funnel. This same concept can be applied to services, which makes up about 80% of any advanced economy. For instance, in our SAP consultancy, we do not have unlimited delivery capability. Nor could we afford to keep enough people on our payroll to deliver an unlimited number of projects.

And here is where you have to understand where all the different elements of your Top Line Revenue[91] comes from. For example, if you go back to our example of having a top line revenue goal of 2.5 million revenue at UP consulting, we can work backwards to the number of consultants we will need. We can do this easily by using what is called a charge-out rate calculator, which tells you, at least for any one individual, given their annual salary plus your company overhead and target profit, what their hourly bill rate has to be.

Then it becomes a modeling question - how many hours at what rate do I need to sell to hit my target profit or revenue goals. As a rule of thumb, within a professional service firm such as ours, you need to sell about 3 to 4 consultants to generate a million in revenue. If you want to earn a million in profit, then you have figure out what your profit margin is and work backwards from there. It's surprising how many people don't realize this. It's also why they are sometimes shocked at the rate their mechanic is charging or their plumber. We are sometimes shocked at how ridiculously low the rates are that Indian based recruitment firms are trying to get Americans to work for. But, hey, they keep on trying and we keep on avoiding them because we've done our math and know what working at a loss looks like. You should too.

---

[91] I use Top Line Revenue here as opposed to Gross Revenue, because Gross Revenue may include revenue from other sources that you may not want to count in this particular process.

Summary and 10 Lessons Learned

Without people, and more specifically, the right people in the right seats, you have no chance of hitting your growth and revenue goals. Automation has and will continue to amplify the productivity of your people, but sales, especially complex sales, continues to rely heavily on people to get the job done. Here's what you need to focus on to solve your HR problem:

- **Right Size Your Team.** Sizing your Sales and Marketing team needs to be based on your overall revenue goal.
- **Clear Sales Plan Required.** You'll need to have a clear sales plan by product or service line.
- **Know Your FLC Rate.** Sizing your Sales Team requires knowing your Fully Loaded Cost for each position.
- **Training and Education is a Powerful Force Multiplier.** Educating sales people is a two-way street - you'll need to provide the training but they also need to actively participate in the training.
- **Master Communication Skills.** There is one key skill all sales people need to master – communications.
- **Multiple Communication Channels Are Always Active.** Communication happens in many forms and venues.
- **Practice Active Listening.** The second key skill or really a parallel key skill is active listening.
- **Personal Productivity Helps Organizational Productivity.** Personal productivity can be taught and improved. You would be surprised how many Account Executives required training on basic skills like how to use an outlook calendar.
- **You Want Outstanding Salespeople.** Good salespeople know their product. Outstanding salespeople know both their product and those of their competitor.
- **A One-Man Sales Team is Dangerous.** You need at least two and

should always have replacements in your recruitment pipeline or better still, in training. It's costly but the only way to ensure your sales.

CHAPTER NINETEEN

The Information Explosion Problem

WHERE DID ALL THIS DATA COME FROM

Now that you have the basics down of setting a Top-Level Revenue Goal, and working backwards to how many contacts, leads and ultimately sales you need to hit them, it's time to talk about the elephant in the room - overwhelming amounts of information. You see, most every business has many systems, some they recognize as systems, and others that are just 'there'.

All of these "systems" are generating data at an incredible pace. For instance, your CRM system is generating data about every activity your sales team is accomplishing. For example, an inside sales team using a CRM such as Hubspot, regardless of product or service being sold, performs five basic tasks. That is send sales emails, call, hold meetings, set up and execute tasks in the background, such as sending out contracts or other sales related documents (think quotes and proposals) or leaving notes (which take time to write and are unstructured data) on deals and contacts.

As it turns out, Sales Activity is a very good predictor (though not perfect) of eventual sales success. The image below is taken from a live customer site, and there's a lot of backstory buried in that chart. The first that should jump out at you is that they went from an extremely low sales activity level of 366 in Sept. 2020 to a peak of 7,430 in July of 2021. In this particular 'mini business case' they hired me and my team the last week of January 2021 as their Hubspot partner.

Let's just say we have an 'extreme' focus on sales success, not just marketing and lead generation.

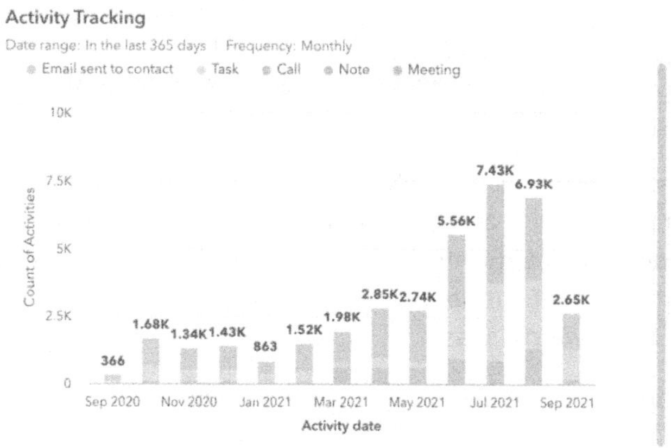

Figure 2. Sales Activity Levels

Sales Follows Activity

Though this client is on the smaller range of your typical successful Inbound Marketing customer, it still is a good fit for Inbound Marketing and Sales because their product is worth more than $500, the lower range of where Inbound Marketing makes financial sense. The astute business owner will, of course, have asked whether all this increase in sales activity resulted in an increase in sales.

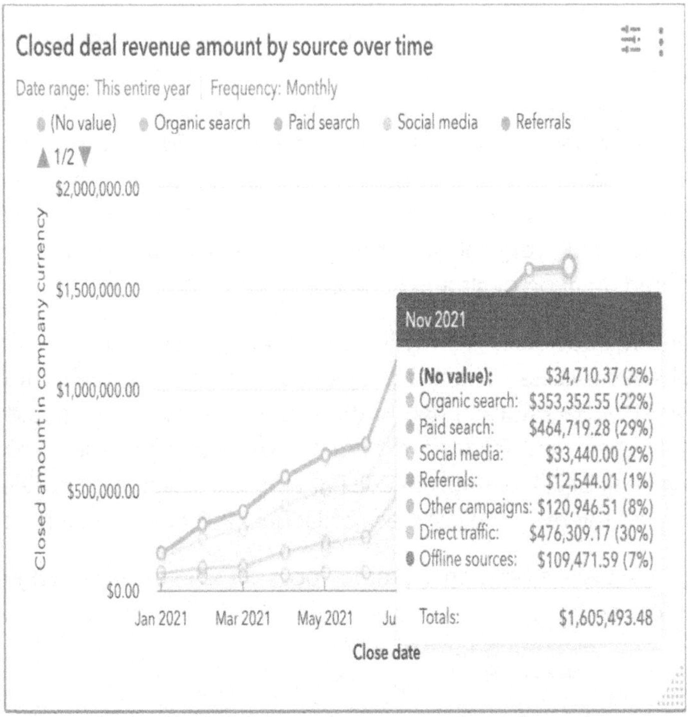

Figure 3. Revenue Results Follow Sales Activity

The short answer is yes. As can be seen in the chart above, it took us a few months to turn it around for them but given that their average sales cycle was almost two months, we eventually got their revenue engine firing on all cylinders. Let me let you in on a secret. When it comes to all that sales activity, all we did was:

- Add one more salesperson to the team (from 1 to 2)
- Make sure they were using the CRM system to assist in their work

That's right. They only had a one person sales team plus the owner (who didn't do much of the selling) when we started, and that person was working largely unsupervised. About the only thing the one person was actually doing was online meetings, when and if someone booked one.

That's the thing about CRM. Sales people generally have not found them to be helpful to closing sales in the past - unless they were using CRMs that were more or less part of the background, collecting the information without effort on the part of the salesperson, who, frankly, is usually overwhelmed with information and just barely keeping his head above water.

Knowing Isn't Managing the Sales Process

Just knowing that all this activity is taking place is helpful, but it provides, at best, directional performance information. You also need to know how effectively the sales team is executing each one of the activities and to be prepared to provide 'sales coaching' to help them improve those activities. After all, if your sales team isn't successful, your business won't be successful. And sales coaching is critical to success.

Therein lies the challenge of the Information Explosion Problem. You see, if you're managing this sales process, you now have between 7000 and 8000 data points to analyze, coming at you every month. For the uninitiated, the temptation will be to dig into every one of these activities and try to 'fix them' which is not wrong, but it will bog you down until the point of utter frustration.

Instead, the trick is to learn what the information is trying to tell you. For instance, the sales activity chart above reveals several aspects of your sales team's performance. Obviously, the low point represented a salesperson who had already checked out of dodge and a distracted management situation, not an unusual situation, but fatal to a small business. Less obvious is the cause of and required action regarding the decline in peak activity.

The chart is telling us several critical pieces of information to go look at. The first of which is whether or not the same number of opportunities have been coming in. The second is whether there is a big change in performance between each of the members of the sales team. If there has been a decline in the number of Inbound Marketing generated leads, then it's time to work on that. If there is a difference in performance between the members of the sales team, then it's time to worry. One of them is about to leave.

If you have a sales mentoring session with them and conclude that a possible departure is in the works, the question then becomes, of course, why? Are they burned out? Is all that sales activity too high of a level of sales activity level to maintain (this can happen, especially when the lead generation is running flat out, which is generally what every marketing team is trying to accomplish). Are they leaving for compensation related issues?

In my experience, this is the primary cause of most departures, especially of top performers. For instance, perhaps you're working in very large organizations, such as I have. You have a quota carrying role, which I have had many times, and you're hitting your numbers and overachieving by 500%, which I have also done. Yet the company holds back 20% of their agreed upon bonus due to the overall 'financials' of the company, you can and should expect a departure.

I've also done projects where very large sales teams, consisting of 100s of members, received their bonus' based on information coming out the datawarehouse system, which the sales team members did not agree with and which they could not audit on their own. This was causing massive sales team turn-over, until we worked out the kinks in the information supply chain.

Critical Few versus Trivial Many

One of the major challenges in managing this information explosion is sorting out what is often referred to as the critical few vs. the trivial many. This is another way of referring to the fact that the information often has a Pareto 80/20[92] pattern to it. For instance, if I have a sales team sending out hundreds of emails, each of which is tailored to the individual who they are sending it to, I don't really have time to read every one of them myself and analyze them. I might want to randomly sample a few of them, just to make sure they are well written.

However, I won't have time nor a need to examine all of them. That's

[92] The Pareto 80/20 rule has been around a long time. It generally means that 80 percent of your results come from 20 percent of your effort.

why I hire competent salespeople and train them. In the case of this particular client, as would be the case for many others, those online meetings are actually one of the primary indicators of eventual inbound revenue, and I would want to sit in on a of them from time-to-time to see how the conversation goes.

The same goes for the calls. Though the system can record calls, as they say, "for training and quality purposes", to actually listen to them requires as much management time as the duration of each call. If the average successful call takes at least six minutes, and they are making 100 calls per person a day[93], that means it would take 10 hours of your time to listen to them. You're not going to go down that road. These recordings mostly serve as protection to the companies, especially with YMYL (Your Money or Your Life) type businesses, but there is gold in them as far as providing sales coaching opportunities.

A smarter strategy might be to listen to calls where you won the deal and a small sampling of the ones you lost. Those might yield useful information that you can use to improve the performance of the other members of your sales team. Should you happen to have interns available, putting them to listening to these could also prove beneficial to their 'development'.

One Bad Apple Spoils the Barrel

That's an old saying you may not know the actual meaning of. What it translates into in the information explosion context is bad data will kill your business. What do I mean by bad data and why is fixing this critical to achieving your growth and revenue targets? Bad data actually comes in a couple of different flavors:

- Data that is incorrect at its source
- Data that is correct but irrelevant to what you're trying to accomplish

Incorrect source data is one of the most difficult to get a handle on. Within the marketing and sales function, one of the most common sources

---

[93] I've used 100 calls a day just to make the math simple. I've worked with companies where the daily call volume was closer to 300 calls a day, and others where it was about 50 to 80.

of incorrect source data is incorrect tracking code installation and setup on your website. Typically, this starts with the Google Analytics tracking code installation. And for those who have migrated to using the Google Tag Manager, the problem only gets more complicated.

It appears relatively simple, you just create a Google 'analytics property' and insert the supplied code on your website, usually a WordPress website. Like magic, if you have any traffic at all, you will start measurement happening on your site. Unfortunately, unless all you want to know is whether or not you have any traffic (many have little to none), there's a near infinite number of little tweaks and settings you have to set up within GA, such as goals, dimensions, metrics, channels, and many, many other settings.

If you have hired anybody in the past to perform any kind of marketing work for you, you may very well find they have installed their own GA code on your site. Google provides a tool, Google Tag Assistant, that will reveal the presence of these duplicate codes as well as a multitude of other errors it normally picks up. Each error usually also comes with a link to Google documentation, that, frankly speaking, is not even developer friendly, let alone marketing, sales or entrepreneur friendly.

Given that 100% of the sites I've worked on are throwing errors, whether they are new, or 20-year-old sites managed by an experienced IT team, you can rest assured that the clickstream coming out of this is less than 100% reliable and is costing you money. How? Simple, GA is normally directly connected to your Google Ads Account and thus, the AI/ML[94] system is receiving less than ideal data from your site for optimal optimization. Unfortunately, it has no idea about the quality of the data it is optimizing on.

That's Just to Start

As we mostly deal with Hubspot hosted sites, which comes with its own tracking code, we always run into the buzz saw of the difference between what GA is tracking and what Hubspot is tracking. I would like to tell you that it is ok to just rely on Hubspot. Unfortunately, if you're spending Google Ad dollars, you have to understand and ensure both measurement systems are working (and talking to each other).

[94] AI/ML or Artificial Intelligence/Machine Learning. Google ads relies heavily on these two very complex technologies.

If your site is 100% on Hubspot, then your tracking is a lot simpler. It's done by Hubspot and optimized for 'Inbound Marketing'. But many, if not most, clients have a mixed environment, and no matter how strong of a business case there is for going 100% Hubspot, many will choose to operate a sub-optimized website.

Facebook, like all other social media sites, also comes with its own tracking code that must be set up. They call it a pixel and at one level, all you do is install it once, and theoretically, you're done. Reality, though, is quite different. You also have to contend with setting up events.

This is not meant to be a Facebook advertising book. But you need to understand that, generally speaking, you will need to set up events as well as install the Facebook pixel. You also need to know that Facebook, like virtually all of the other systems that ask you to install a tracking code, makes constant changes to these codes. You also need to be aware that Facebook owns many other properties you are likely to also be using, such as Instagram, which also have their own unique tagging requirements.

In one of the more complex environments I worked in, we had dedicated people to each social media channel, and also did a lot of television advertising. Each of those channels, i.e., Snapchat, Pinterest, Twitter, to name just one of 13 different platforms we maintained a presence on, all had unique tracking codes that needed to be installed.

Each of these social media channels generated millions of data points and each actually is focused on different aspects of their users. That's why Hubspot is such a powerful tool, it helps bring the social media data flows into Hubspot and helps you sort the wheat from the chaff.

Internal Data Sources also Exploding

I've mostly focused on the challenges of acquiring and managing the flow of externally generated data from various systems and platforms. However, they are just one source of a vast flow of information you have to wrangle. The other source is all of your internal systems. For instance, you most likely have some sort of ERP or Enterprise Resource Planning system running. You may not call it that or even recognize that it is one. However, if you sell anything that involves any kind of inventory, or even if it does not, you will have need of a system to manage inventory, manufacturing, warehousing, financial management and human resources management.

In fact, (and putting my SAP and Oracle hat back on), you probably will come close to needing 16 different 'modules' to run your business. And every one of them produces a torrent of information just from operating. All of them also, directly and indirectly, have to integrate with your sales and marketing systems.

And all of these highly integrated processes, just by operating in an integrated fashion, generate massive amounts of data about the data they are generating. For instance, if you run an ecommerce operation, you may have outsourced your manufacturing operations overseas, which is a very common scenario. You will find you need complete inventory visibility in order to satisfy today's picky consumer and their 'got to have it now' behavior.

Beyond knowing what your inventory is or will be, you also need to be able to predict future demand and reorder optimal amounts of inventory. That means your ecommerce system needs to not only talk to your inventory system, it also needs to be driving your predictive analytics system. That sounds like space that only the largest of enterprises play in, but it is now possible that even one-man shows can and are doing this.

Though this is an 'internal' data source, in our scenario above, there is also an implied connection to external systems, in this case, to the outsourced manufacturing operation. It is almost a guarantee that they won't be using the same system as you do. This usually means an integration technology layer is required in order to facilitate this type of multi-enterprise collaboration. This used to be almost impossible but today, it is far easier with the rise of cloud based integration applications, such as Zapier[95].

---

[95] Zapier is a cloud-based integration platform that allows you to connect two systems and exchange data. There are other systems as well in this category, and there are other much more complex enterprise applications as well.

Summary and 10 Lessons Learned

Information is called the new oil of the information economy. Every business generates data from a variety of sources, which all need to be assembled, often in real-time, to yield valuable sales and marketing information. The lessons learned about the information explosion:

- **Tracking Sales Activity Generates Massive Amounts of Data.** From a sales perspective, the ability to track activity is a key information explosion generator.
- **Be Judicious in Your Data Tracking.** There's a fine line you must learn to draw when deciding which sales activity information to examine and possibly drill down into.
- **Sales Activity Indicator of Future Sales Outcomes.** Though not directly attributable, sales activity is often a very good indicator of future sales success.
- **Analyze for Insight.** Because it is trackable you can usually work out which activities or combination of activities result in a sale.
- **Set Activity Goals.** It is also possible and necessary to set activity goals. Ideally, your CRM will provide the ability to track these activities automatically.
- **Use Alerts Effectively.** Most CRM systems allow you to set up alerts for activities that are outside expected limits, either up or down.
- **Understand the Challenge of Setting Up Analytics.** Getting your website's analytical tracking codes setup from all of the different sources can be a challenge. Though we believe Google Tag Manager (GTM) is a clear leader in tag management, there are many other tag management solutions available, and may serve your purposes better than GTM.
- **Tweaking Tracking Codes Never Finishes.** These tracking codes require fairly constant tweaking as they are constantly updated by their suppliers, such as Google.
- **Integrate With Your Suppliers.** You need to connect your systems

with your supplier's systems, which presents a whole other layer of complexity.

- **Critical Few vs.. Trivial Many.** Internal data sources can generate overwhelming quantities of information. Only a small of portion of which is useful from a decision-making perspective. Spend time figuring out what to listen to and what to ignore.

CHAPTER TWENTY

Marketing Tool Miasma

WHY MARKETING IS KEY

I've previously walked you through the process of figuring out how many leads marketing will need to generate so that sales can hit its revenue goals. However, I have not really laid out the challenge you will have getting marketing to be a source of profit versus being a cost center.

Need an Ideal Marketing Platform

I've already mentioned that I am a Hubspot Inbound Marketer in this and other books in the series. If you have already worked through the math of setting a definite revenue target and working your way backwards to the number of leads required, you're way ahead of most businesses.

Though I am big fan of Hubspot, it, unfortunately, will not solve all your marketing needs. I do consider it close to being an ideal marketing platform. It still has some weird and frustrating limitations after all these years, like not having built in ecommerce capabilities. But you can always get that capability, at an additional cost, from a variety of other solution providers.

And that's the marketing problem. You're almost guaranteed to need not one or two different systems, but rather, you will most likely end up with 8 to 10 systems or even more. Large corporations may well have over a hundred. In fact, there's a map of the marketing and sales system landscape out there that shows there are at least 8,000 such systems available, and that number grows every year.

Many of these systems are very good and might actually be useful to you hitting your growth and revenue goals. The challenge is for you to figure out what you need and to spend the time evaluating them, purchasing them and integrating them into your systems landscape. The marketing departments of many companies now spend far more on IT than the IT department does. Because CIOs are, by design, conservative when it comes to new systems acquisition, marketing long ago started to go around them.

We Don't Evaluate Systems for Free

As Hubspot partners, it would be tempting to just do Hubspot. Most clients eventually figure out they want or need more in the way of systems and turn to us for system recommendations. Initially, we could take this task on, but experience has shown us that unless it is a formal, paid for activity, it is mostly a waste of our time.

Why is that? Simple, it can take days and weeks for us to research a client's requirements[96], come to a conclusion about what might be needed or is likely to work, and then survey the market for potential solutions. For instance, one of our clients asked us to evaluate ecommerce systems. That sounds simple enough, but until you really, truly understand how a business operates, in other words, until they know what they sell and how they sell it[97], it is impossible to make an objective evaluation. In the world of small enterprises, who tend to move very fast, this formal evaluation process isn't often done to the level it needs to be and many systems are tried and abandoned after they prove a poor fit.

That's why I am glad to help guide a client to evaluate the certified applications available on the Hubspot marketplace. Hubspot has at least validated they talk to Hubspot. But they have not said what it will cost to make use of the apps. For instance, let's say a client wants to have a pricing calculator[98] on a page somewhere. The process for building that can take quite a while. It usually starts by building it in excel first and then, building it in the calculator builder tool you've just subscribed too. But then you also have to build it a custom landing page for it to live on, and integrate the individual calculator fields[99] to the corresponding Hubspot fields (if they exist, otherwise, you also have to create those as well).

---

[96] This would be at a minimum. I've seen the Big 4 consultancies spend up to three years on a client's site gathering requirements, and of course, those were all senior level analyst positions. They don't come cheap. But realistically, most companies will not, nor should they have, these resources available internally.

[97] I wrote a book on just this subject, available on Amazon. It is called Knowing What You Sell: The Key to Dominating Your Market, https://www.amazon.com/KNOWING-WHAT-YOU-SELL-DOMINATE-ebook/dp/B0946YYDK2/ref=sr_1_3?dchild=1&keywords=knowing+what+you+sell&qid=1633606503&sr=8-3

[98] Calculators in general are great lead generation tools. However, there are a multitude of tools available to build them, all with features and benefits that are very distinct.

[99] The reason you want all of the individual fields integrated into Hubspot is because then you can use these fields to generate contextually relevant emails and even proposals that use these fields as personalization tokens. At the enterprise level of Hubspot, you can also go directly from these fields to calculated fields that, for example, allow you to generate customized quotes based on the calculator inputs.

After that, you still need to create what are called lead nurturing emails driven by lists and workflows. Finally, you most likely have to incorporate this new piece of content into the sales team messaging scripts. Once all this is done, it will begin generating leads, you hope, and it will throw off a lot of new information, which you will invariably want to analyze, taking up yet more time.

This all requires the development of an overall campaign strategy. That's why this is both a technical task, i.e., designing, developing and deploying the calculator and marketing strategy task. This is where an expert Inbound Marketing expert is worth their weight in gold. While some clients may have the internal ability to come up with this campaign strategy, most do not, and therefore, need some external help to come up with the overall marketing and sales strategy design.

Interactive Content

Though a calculator is technically interactive content, specifically, it is User Generated Content (UGC), it provides a fairly low level of interactivity. These days, most prospects have very high-level expectations of your content, and are looking for 'experiences', meaning highly interactive content. You can either develop this from scratch, a daunting task, or you can use one of the many platforms out there for developing this, i.e., SnapEngage®, Ceros® and others.

They are very expensive and usually require a great deal of thought and effort to use. If you get it right, they can become your lead generation gold mine. But most of the time, their payoff is mixed. It just depends on what you sell and how well you can turn it into an experience. I highly recommend interactive content and expect it to continue to grow in importance. However, I also expect smaller or even mid-size enterprises to struggle to fund it.

Marketing is Very Broad

Here's where the biggest marketing problem comes from - it is huge. Most marketers fresh out of college have no idea that they will need to be both creative and have to learn how to set up analytics. The range of what is expected from the marketing team is vast and ever expanding. It is also extremely analytic in nature, and though you don't need advanced statistical math skills to succeed in it (but they certainly help), you do need solid math skills and will find that you're always doing some sort of calculation.

The boundary of where marketing ends and some other function begins is not only ill defined, it is also not easily definable. What do I mean by 'ill defined'? Here's a typical scenario based on working with a number of my clients. You have setup your marketing system and are generating leads, in both the quantity and of the quality demanded. In short, marketing has delivered. Unfortunately, sales have not kept up with lead generation efforts. In 100% of the cases, sales will immediately say the quality of the leads is no good. You've heard the line, Glen Gary was not a fictional movie, it was more of documentary...

The owner typically immediately will demand action be taken by marketing as clearly, they've done something wrong. However, the smart marketer will usually then find themselves deeply interacting with the sales team. Sales is tough but most businesses bend over backwards to make it/them work. However, with a system like Hubspot CRM, where literally every action sales undertakes, i.e., calls, meetings, emails, tasks and even face-to-face meetings, can be measured, you will usually find out something is amiss. For example, perhaps the measured sales activity level has dropped, as in the example below.

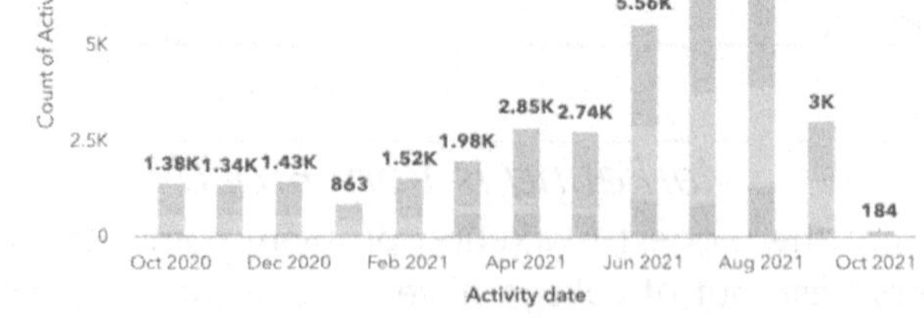

Figure 4. Sales Activity Drop

That's an actual screen shot from a live customer with a tiny sales team. They work exclusively in an inside sales environment. There are only two of them typically. What should jump out is the pattern of activity. It was at a low but steady rate until Feb. 2021 but had previously collapsed by Jan. 2020 (not shown on the graph). In this case, me and my team were hired to 'fix it' in Feb. 2021. It took a while, but after we got the plumbing sorted out, meaning, everything connected and all systems functioning: phone, email, meeting links, etc., and replaced the existing sales team, we saw a slow but steady climb in sales activity. As it happens, revenue increased in exactly the same pattern.

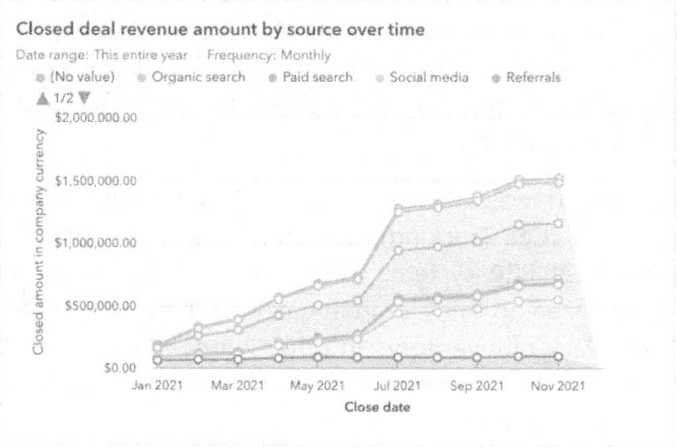

Figure 5. Activity Steadily Increased

Though this is a very small business, I've seen the same results with much, much larger businesses of every description. As sales activity increases, as sure as day follows night, sales increases. But why tell you all this in the Marketing Problem chapter? It is simply to point out that marketing is usually involved in some or all aspects of sales. But it isn't just sales where marketing ends up spending a lot of time.

Logistics is also usually a part of the organization where marketing gets pulled in. This is especially true in smaller enterprises, though it happens in every size enterprise. For example, one of our smaller ecommerce customers kept running out of product. The fix demanded by the owner was for marketing to fix it. Most marketers won't have the same logistics and in particular, Supply Chain Management background, that my team and

I have. But, fortunately for this business owner, we had the requisite skillsets to both assess and diagnose the issue, using excel and Industrial Engineering expertise. We eventually found an application within the Shopify eco-system that could handle the issue in a more automated, if limited fashion. That's how marketing comes to be poking around Reorder Point Planning and Demand Planning.

In my experience, this interaction between marketing and everybody else in the organization happens no matter the size of the enterprise. The only real difference is how direct the interaction is. For instance, in one of my mid-size clients, doing about $50,000,000 a year, they actually had a formal IT organization, with 3 or 4 members on staff and a slew of contractors and consultants on call. Brought in as their Hubspot consultant, to focus on 'fixing' Hubspot, whatever that meant, I found myself in-charge of about 13 members of the marketing team. I was brought in, of course, when their revenues had collapsed down to about $20,000 a month, and things were looking grim.

What I found was a very complex IT landscape, including Hubspot, Salesforce, Five9 (Call Tracking), Quickbooks, Zuora, Google Business Suite and a home-built ERP system. After an initial period of doing 'Inbound Marketing' and seeing fairly dramatic results, I started to get the message that we had a lack of inventory. In other words, slow down. Unfortunately, that's not really how Inbound works. It works whether you want it to or not, once you have it working.

Track Every Contact to Customer Event

The reason I mention all these systems is because you need to be able to track every opportunity through your system. That means from the time some anonymous visitor hits your website or walks into your store until they become a paying customer. That may sound like an impossible task, but with today's digital technology, it is entirely possible.

If you're asking for paid advertising dollars, it is entirely the norm that the CFO will have the unreasonable request to be able to track every 'click' (which is what you're usually paying for, in one way or another) all the way through to a revenue. Click-to-dollar is the basic task.

The ability to do this relies on having an integrated suite of systems, across your enterprise. That means Finance and Logistics and Human

Resources are all talking to each other at a deep, systems level. It also means that the strategy management system is talking to and is being continuously fed information from all of the systems, typically via a Datawarehouse.

Your Leads Do Have a Computable Value

From a marketing and sales system perspective, it all boils down to knowing that every lead has an actual value, which can be computed. I've covered how to calculate this in my previous books and won't rehash it here. However, the key takeaway here is that knowing that every lead has a value, means that everybody in the organization, regardless of which system(s) they might use, also has an impact on capturing that lead value.

Ensure All Systems Integrate with Hubspot

Though this book is not exclusively about Hubspot, it is probably useful to think of Hubspot as 'whichever' marketing platform you use. In the case of using Hubspot as your central marketing platform, whenever marketing wants to bring another system onboard, the first question you need to ask is whether the system is 'natively' integrated with Hubspot, for instance, like Salesforce is, or whether it can be integrated 'easily' using something like an API.

This concept of making sure all your marketing or sales systems (CRM) are integrated isn't unique to a Hubspot centric marketing organization. It's true whether you use Marketo, Pardot, or any of the dozens of other marketing platforms out there. None of them are able to meet all of your needs, which means you will need other systems or will end up extending the system's functionality via custom development. What you do not want is to be forced into hiring staff to keep them talking to each other.

What About Logistics?

You might be wondering if or whether your ERP system needs to be integrated with your marketing platform? Short answer, yes. There are many complex scenarios where an individual name needs to be associated with a product. That's well known. However, there's an even greater challenge most organizations face, that is tying 'demand creation' or marketing campaigns to product or service availability.

There's no greater waste than having your marketing team generating demand that you have no capacity to fill. This gets extraordinarily complicated to manage if you are working in siloed information systems environments, which I have often witnessed.

Finance also needs to be tightly integrated into the marketing system, though it doesn't always appear to be so. One of the key 'pain points' is when you're running large Google Ad-Spend campaigns. The smaller the enterprise, the more noise this generates when it comes to finance. Depending on your level of spend, you will want to get your Google Ad-Spend (this holds true across all platforms where you're likely to be spending money) on a monthly invoice scheme with spending alert levels built in. They have somewhat clunky mechanisms for controlling overall spend. Usually in their favor.

Summary and 10 Lessons Learned

Marketing and Sales has become extremely complex, which has driven the rise in the number of systems the marketing team is using. At the same time, business has come to rely on all of their systems being fully integrated, at a deep systems level. The marketing system lessons learned:

- **Simplicity Without Limiting.** Requires that you have the fewest systems possible and yet as many as necessary.
- **Ecommerce Requires Deep Knowledge of Your Business.** In order to evaluate ecommerce systems, you need a deep understanding of how the business really operates, versus how it may believe it operates.
- **Interactivity Drives Engagement.** Content interactivity is the expected norm. Dreaming it up, developing it, and deploying it requires a combination of systems and skills that are expensive.
- **Creative People Required.** Creating interactive content can require expensive systems coupled with highly creative people.
- **Moderate the Sales vs.. Marketing Conflict Points.** Watch out for sales issues driving changes to marketing, it is often counterproductive and costly. If you can solve it with data, you're much more likely to win.
- **Typical Quick Win for Many Businesses.** For many businesses, an online quoting calculator can quickly help generate a high number of leads. Don't make it hard to sell by making it hard to buy.
- **Make Sure You've Got Activity Tracking Up and Running.** A good CRM provides the ability to track, in the background, almost all sales activity.
- **View Your Customer's Journey-to-Sale.** With the right analytics in place, you can track every activity a prospect took before becoming a customer. You can also use this insight to uncover points of friction so you can work to eliminate them.
- **Look For Native Integration but Don't Limit Systems Choices If**

Possible. When you run a marketing platform like Hubspot, it is critical that any new system you acquire is integrated with Hubspot, preferably natively.

- **Have a Measurable Process Driven Enterprise.** Every system you bring onboard will have an impact on many processes. Be aware of these processes and be prepared to make course corrections.

CHAPTER TWENTY-ONE

Get Your Budget Problem Under Control

IT'S GOING TO COST HOW MUCH?

I've shown you how to set a sales goal and work through the process of setting lead generation targets and customer acquisition goals. I've also explained why this all requires systems of various flavors to make work. Now it's time to figure out just what it is going to cost to achieve these growth and revenue goals.

Rules of Thumb

I hate to break it to you, but information may want to be free, but systems aren't free. You're going to need systems for every part of your organization, and for larger organizations, like the United Nations or Saudi Aramco (both old customers of mine), which have a million users or more, the cost of their systems are measured in the tens-of-millions of dollars every year. And this is typically some small fraction of their annual revenue or budget.

For most businesses beyond the early start-up stage, using a percent of revenue figure is perfectly feasible although not very scientific. As I want to focus just on Marketing and Sales, let's look at the foundational number: Contacts.

More Contacts Equal More Costs

In general, but not exclusively, most marketing platforms charge a combination of monthly subscription fee plus a cost per 100 or 1000 leads stored. This makes sense, as the purpose of the marketing system is lead acquisition, yet on the sales side, it's a little more complicated as they are typically focused strictly on revenue generation.

Many customers want to 'delete' contacts once they determine they will not convert into customers. This seems reasonable on the surface, but then you run into the whole GDPR[100] and other related privacy laws that leave you exposed to fines for not managing people's data properly.

---

[100] GDPR - General Data Protection Regulation. This is an EU regulation theoretically designed to protect individual data. It also opens the door to massive lawsuits and is one of the most dangerous business regulations ever promulgated.

That's why I recommend you NOT delete contacts. I've already seen how very old, very litigious contacts[101] can come back to life and cause a great deal of trouble because a client did not properly control his emails or phone calls or SMS activities. That means, unfortunately, you probably need to keep those contacts in the system for the life of the system and the business. However, that still won't excuse you from also practicing comprehensive email hygiene, meaning keeping track of unsubscribes and other activities so you can prove it when it needed.

Pricing Changes Constantly - Usually Upwards

As I previously mentioned, you're going to end up with a lot of different IT systems, each of which has its own particular pricing model. That's why I am not going to provide pricing for any particular system. However, I can tell you that you need to create a system landscape directory and put them in there. Your finance guy will appreciate it. More importantly, you can figure out what you're paying to warehouse contacts, which is typically not much, but it can add up, especially if you have millions of them, which many of my clients do. But because they have such large numbers of contacts, they also get a nicely negotiated deal from the system providers.

Don't Let Cost Keep You from Success

When it comes to your marketing team, they are going to want you to have access ($$$) to different systems. The general tendency is to say no, there's no budget. But hear me out - the right answer depends entirely on what you're willing and able to pay for a lead any how many leads it takes to generate one customer. If the new system can help move the needle on either one of the variables, then it's just a math problem, followed by a budgeting exercise.

Remember, if you have smart people on your team who take the time to look around and evaluate such solutions for you, and you just say no, they won't keep looking around for you. Exactly zero procurement departments will go looking around for solutions for marketing departments.

---

[101] I have customers who routinely have to turn over a copy of every bit of data they have to authorities for 'investigation' due to a customer complaint. Though the customer has never actually had to pay a fine, the cost of compliance is enormous, running into the millions of dollars on annual basis.

Evaluating a System's Cost Using the .001% Rule

Over the years, I've become what you might call a lazy value engineer[102]. You see, I happen to be an SAP Certified Value Engineer. That means, that in one role, I am able to perform complex business case development exercises for very large complex systems such as SAP as part of the sales process. But these exercises are typically very complex, fairly involved, manpower intense affairs. For most of my clients on the marketing side, I use a short cut. The .001% shortcut.

How does it work. Very simple. I ask a simple question. How much would a 1 tenth of 1 percent improvement in whatever it is I am trying to do, improve yield? For example, my value engineering team and I developed a calculator for one of my clients in the warranty management space. The client's software is used to improve the complete warranty management process. If you look at a publicly traded manufacturer, like General Motors, you will find they have a line on their income statement for warranty expenses and Gross Income. As it turns out, a one-tenth of one percent improvement in the warranty expense ratio for GM is a massive number. The calculation is very simple and is broadly applicable to any company that has warranty expenses they need to deal with.

Using this same concept, you can evaluate ANY marketing or sales system. You have two routes to system evaluation simplification. Either focus on COGS (generally, you want to decrease this number if you buy a new system or methodology, and this is on your financial statements. Or you can use (provided you have analytics up and running) your existing lead-to-sale conversion rate.

You can either do the cost justification using Visits-to-Leads or Leads-to-Customers or Visits-to-Customers rate. Of course, you need sufficient volume for the math to make sense. However, if we take just one of our customers, where we are generating about 10,000 contacts a month and 2,000 customers a month with an AOV of $280, we have all we need to do a quick calculation of how much improving contact generation by just one tenth of one percent will yield. In this case, that would be 10 additional leads per month and given the conversion rate (20%), 2 new customers or about $560 in revenue, per month or $6,720 per year.

---

[102] A value engineer is actually a term first used by the United States Government way back in the 1950s. The role has many flavors, but generally falls into what is called Industrial Economics.

This is about the smallest unit of change I like to deal with when 'pitching' a change to a client. It tells you, based on your break-even rate, what you can afford to spend on a new system. It is not saying you will only get .001%. It is, instead, providing you with a what-if tool. Now you know, if you improve traffic, all other things remaining unchanged, you can expect x amount of revenue. If you improve your conversion rate, you can compute exactly how much additional revenue that improvement will yield.

Summary and 10 Lessons Learned

Marketing and Sales Systems cost money, a lot of money. You can expect these costs to continue to rise as more and more of the providers realize what it really cost to hit an acceptable MRR rate[103]. What you need to keep in mind about budgeting for these systems:

- **Price Transparency is Critical.** Most systems have straight forward pricing available right on their website.
- **Use Your Negotiation Power.** You can generally negotiate some discount if you are buying 'big enough'. For smaller subscriptions, it's usually not cost effective.
- **Understand Your System Vendors Price Metrics.** Many if not most marketing systems charge by the number of contacts you have in your system.
- **Understand What Constitutes a Contact.** Some only focus on emails, while others don't require an email. Others go either way.
- **Be Aware of the Impact of Deleting Dead Contacts.** I don't recommend you delete dead contacts. They tend to come back to life. Instead, implement good email hygiene practices, like suppression list based on failure to open emails.
- **Compliance Issues Must Be Addressed.** For YMYL (Your Money or Your Life) businesses, being able to track the complete lifecycle of a contact is critical for compliance issues.
- **Enhance SMS With Behavioral Information.** If you're one of those outfits driving huge numbers of outbound SMS messages, you'll get better results with an Inbound Approach rather than with a purely bought leads approach.
- **Combine Paid Lead Generation Approaches with Inbound Approaches.** However, you're unlikely to generate the desired lead-volume with purely Inbound Lead Generation approaches without paid promotion.

---

[103] MRR-Monthly Recurring Revenue. Most subscription-based businesses are based on this type of model.

- **You Need an IT System Landscape Map.** You need to develop a comprehensive system landscape model, so you can keep track of what you're paying for and track price changes over time.
- **Be Aware of Who 'Owns' Your Systems.** You also need to keep track of who bought the system and what it is being used for. Often times, the original purchaser leaves the company, and the purpose is no longer relevant.

CHAPTER TWENTY-TWO

Multi-System Patchwork and Plug-In Mess

IT'LL NEVER FLY

Did you know that about 1/3rd of all websites run on WordPress? That's a huge number of websites. Do you have any idea of how many WordPress plug-ins there are? Or even what a plug-in is? Turns out, there at least 37,000 plug-ins available and as many as 50,000 plug-ins, depending on which directory you look in.

You can do almost anything you can think of with these plug-ins. New ones are being built every day to do one-off tasks. However, there's a problem; most of these plug-ins are either poorly supported or not supported at all. This is especially true of the 'free' versions. However, even when you pay for a plug-in, you may find the support is less than stellar to downright crappy. Yet, many businesses run on a WordPress website, often referred to as a Frankensite[104].

Has An Impact On The Bottom Line

One of the 'joys' of running a WordPress website, especially if you're a business owner whose website is core to meeting his growth and revenue goals, are plug-in updates. You see, every plug-in has to be routinely updated. Yes, this can be set to automatically happen. No, that's a very bad idea. As I indicated, many of these plug-ins are very poorly supported. They most definitely do not have enterprise class configuration control going on within a global development environment, like an SAP or Oracle system has. But they don't cost too much.

Here's the issue you have to be aware of when it comes to those updates - they break things. All the time. Therefore, you must either have the in-house capability to find and fix the problem, or you must have the budget to have external support available. And as usual, the smaller the business enterprise, the more likely time is to be wasted on 'negotiating' the price of these fixes, costing you money in terms of lost revenue and essentially preventing you from reaching your revenue goals.

The average WordPress website seems to have about 20+ plug-ins in it, while some have a hundred or more. The problem always comes down to

---

[104] Hubspot vs WordPress Comparison: https://www.hubspot.com/comparisons/WordPress-vs-hubspot

support. It's one of the primary arguments for having a platform like Hubspot. They have a huge team of engineers who are dedicated to fixing their platform and do a good job of it. This is also true of Salesforce and to a lesser extent, SAP[105] and Oracle[106].

Not Very Secure

Your site is under attack. Guaranteed. All the time, no matter what you do. Don't believe me. Check out this screen shot from one of our clients who runs a security system (at quite a bit of extra cost) on his WordPress website.

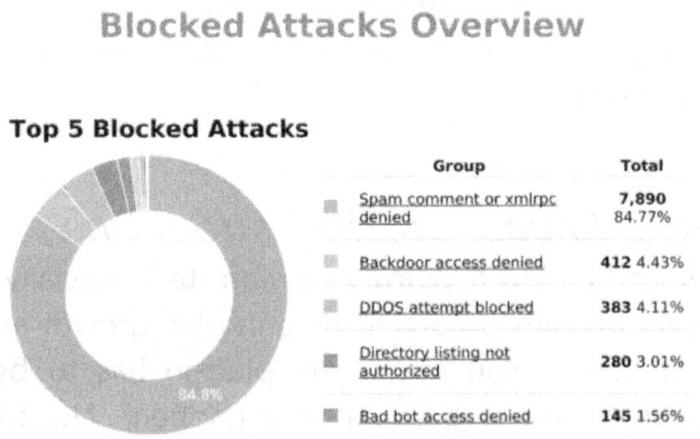

Figure 6. Blocked Website Attacks

This is not a heavily trafficked website, yet is responsible for millions in revenue for this small client. This is for about a month's worth of traffic. You will notice that there isn't one type of attack going on, there are many, and in large numbers. That shouldn't surprise you. What should surprise

105 SAP actually has an enormous customer support team. It is funded by the annual maintenance fee. While support often gets a bad rap, having managed it for the Middle East for a period of time, I can say we did a very good job of fixing problems. The challenge is the sheer number of systems and possible problems one must be able to address.

106 I am sure Oracle now has a large team of support people as well. However, when I worked with Oracle in the 1990s, in the Middle East, support consisted of a terminal that they would let me set down at in their office and do my own ticket research.

you, if not frighten you into action, is that this security isn't an out-of-the-box capability of WordPress, but is available for purchase. What you may not appreciate is that there is no level of security defined for this website. You see, within the cybersecurity space, there are highly defined security level protocols, and a system like Hubspot (and others, such as SAP) have publicly stated[107] they meet those protocols.

They are the ones making sure all of their customer's sites remain secure. When they experience an attack, they have a dedicated team who take on the task of repelling it, so you don't have to. For the smaller enterprise, who, as you can see above, is under constant attack, this is a huge benefit to the platform. You see, the attacks will never cease. The war will never come to an end. But you can't opt out of it either. It is never the case that the only way to win is not to play the game.

Every Plug-In is a Potential Security Vulnerability

All of those WordPress Plug-ins, as useful as they are, and as cheap as they are, represent a security vulnerability that can be and is routinely exploited by bad actors. How risky are they? Let's just say the business that the above screen shot comes from, after having their website go down for over a month, learned the lesson the hard way.

---

107 https://www.hubspot.com/hubfs/security_documents/HubSpot_Security_Overview.pdf

Summary and 10 Lessons Learned

Just about all businesses of almost any size rely on one or more systems, which were often bought without a thorough system analysis and requirements design phase. Even when an enterprise does perform this task, they still end up with a multitude of systems. Each represents both a potential single point of failure (SPOF) and a security vulnerability. All websites are and will continue to be under constant attack by bad actors. Though this book is about growing your revenue, I believe it is critical to take a systems approach to ensuring you hit your targets. That's why I've included this chapter on IT systems. Almost all modern sales teams rely heavily on a variety of systems for every phase of their sales process. Get this wrong and sales will suffer. What you must know and do to deal with this reality:

- **Map Them Out.** First, identify all the systems you actually use. It doesn't have to be a highly technical exercise, but it does have to be 100% comprehensive.
- **Assign Ownership Responsibility.** Once identified, find out who owns those systems and why the businesses needed them when they were first procured.
- **Validate the "Why" Of a System.** Verify the reason still exist for 'owning' them. If no reason can be validated, shut them down.
- **Be Paranoid About Security.** Evaluate the systems from a security vulnerability standpoint.
- **You're Always Subject to Attack.** Assume every system either is compromised or could be. Many security compromises are internally driven.
- **Pick a Secure Website Platform.** If running a website like WordPress, ensure you have defined what security you need and whether you truly can trust your operation to such a setup or would be better served by upgrading to a more robust, highly secure platform, such as Hubspot.
- **Keep Your User Database Clean and Accurate.** Routinely evaluate

the users who you have on your system[108]. Do they still need access to the system?

- **Plugins Must Be Secured.** If you use plugins of any type, assume they are less secure than you need.

- **Be Aware of Home Brewed Plug-ins.** If you build your own plug-ins, ask who and how they will be maintained over time. There's a lot of turn-over within the IT space, not all of them have the same capabilities.

- **Manual Integration Maintenance Drives Cost Up.** Look for opportunities to reduce or eliminate manual integration or non-native integration points. All internal system processes, such as the procurement process within your ERP system, should be natively integrated and maintained by the vendor. Any subsequent upgrades that induce issues with these integrations are the responsibility of the vendor. Thus, very small vendors are at a distinct disadvantage when it comes to selling into the enterprise space.

[108] You will frequently hear this referred to, among other things, as SOD or Segregation of Duties. That is really just a matrix that defines what User Roles and Responsibilities a person will have access to.

CHAPTER TWENTY-THREE

Get Your Sales Team Onboard

SALES AIN'T WHAT IT USED TO BE

Forming, Norming, Storming and Performing[109]. You may have heard this phrase before. I think it is a pretty handy meme when it comes to standing-up, training, evaluating and coaching a Sales Team.

Get Your Sales Team Sized Right

Many products and services require an actual human being to make the sale. Many of these sales transactions may happen in a system like Shopify or WooCommerce, but that is just how the transaction is handled. The sale is still two people communicating with each other, until one either says yes or no.

That communicating part takes time, a lot of time. Yes, there are easy formulas for figuring out how many people you will need given certain parameters. For example, one of our clients has a large Call-Center staffed with about 130 agents. Our figures show that to make a sale, the initial conversation typically last around six minutes. Since they work for 8 hours a day and need to take a break every couple of hours, usually for about 20 minute each time, it means there are really 7 hours and 20 minutes of available Agent time to make those calls. If you do the math, that means they should be doing about 73 sales a day. But the real number is probably closer to 13.

Within this call center, variation between the highest closer and lowest closer is vast. This is due to skills and motivation primarily. However, if you ask any sales person there, they will usually say it is the quality of the leads. I have a slightly different take on the issue.

It's Your Offer

There's a famous Internet Marketing Guru, named Perry Marshall, who nailed the issue. It's the offer. If there is one thing you can do to get higher performance from your sales team, it is to improve your offer. That doesn't mean just make it cheaper (though it might mean that in the right circumstances). It means improve every element of your offer, from how you market it to how you price it, to how you sell and service it.

---

[109] The concept first appeared in a document by psychologist Bruce Tuchman

Let's explore a couple of examples, one good and one bad. The first one is from the diet and fitness industry, one of the most difficult, yet lucrative industries out there. In this case, our customer had a proven offer and had previously been making steady sales. Then sales collapsed. From 2 to 3 million a month down to 20k a month and heading toward zero. What went wrong.

The first major glitch was their shopping cart. They had built an ecommerce platform entirely within Hubspot, which was very unique at the time. However, the conversion rate had steadily begun to drop. The product was the same, the number of inbound leads was the same. The call center was the same. But something had broken.

We began by testing the shopping cart - by making actual purchases and receiving the product at our house. Like all good Time and Motion Studies guys, I measured every step of the process. I found many little problems with it. The initial measured conversion rate was around 1%[110], not good.

The first thing I noticed was that the offer was very confusing. You had to work very hard as a consumer to figure out exactly what you were buying. I fixed that, and the conversion rate improved to 3%. Then I compared our shopping cart to several other sites selling similar products. On each, I found little tweaks we could incorporate, such as more and better 'Trust Seals' and other indicators of trust. This got us to around 6%. After that, we made a series of ongoing Conversion Rate Optimization focused changes that eventually got us to a 13% conversion rate - a 13X improvement from the initial rate. All of these actions fall under what I consider improving our offer.

Then we started improving the marketing materials, to be much more educational in nature. We already had strong lead flow, approximately one conversion every minute, on a typical day, which is very good for most businesses of this size. This allowed us to indirectly improve the number of subscribers, which dramatically improved cash flow. Finally, we got our product on Amazon and into the Fulfillment By Amazon[111] program, which dramatically improved revenues.

---

110 I call this 'not good' for this customer. There are plenty of stores for which this is about normal.

111 The consumer knows this as Amazon Prime.

Now for the Bad One

For the bad one, let's focus on the tough one, ERP. You see, selling complex software, such as SAP software, is a time consuming, complex undertaking. Doesn't mean it can't be done, but it does mean it takes a lot more resources, in terms of time and effort, to seal the deal.

One of the toughest aspects of selling it is the shear complexity of the offer. You see, implementing a complex system such as SAP is akin to changing the motor on a fighter jet, while it is in flight and engaged in a dog fight.

Before you get to an implementation though, you have to sell it. And that is where improving the offer is the challenge. I should know, I worked as a quota carrying SAP Industry Principal[112] and had to put together custom offers all the time for very complex solution designs, for such entities as Airlines, Railroads, Oil Companies, Telephone Companies, Manufacturing Companies, Construction Companies and Professional Service Companies, to name just a few.

One of the challenges with the offer was what was called the Quote Tool. It consisted of a very large, complex, black box, excel based tool which you used to select and price out the software your client would need. Of course, this meant you (or I) had to not only understand the functional and technical requirements of the customer but as well, had to be able to figure out which software components were going to be required to meet those requirements. Did I mention there were over 1,300 software components and other elements on the price sheet (or Quote Tool)?

If you got this wrong, which happened, eventually, during implementation, somebody was going to have to go back to the client and inform them they needed to buy additional 'stuff'. You can imagine how this went over. Reasonable clients understood this situation and were often willing to 'go along' to move things forward. Unreasonable clients[113] brought lawyers to the table.

I always thought this was the worst possible way to do this. Especially since we had a completely different model available, but only to our largest

---

[112] I also worked as an SAP Program/Project Manager, and ran Professional Services for SAP in the Middle East. All of these roles also had a revenue generation element to them.

[113] Almost all clients fall into this category, especially within the enterprise space.

customers - the "all you can eat model" for enterprise customers. Under this model, someone like me would still have to figure out which components were going to be needed, but the price was fixed, at so much a year. If I missed something, all I had to do was amend the contract (at no cost to the customer) and the missing software component would be provided to the client.

This was a far simpler model for the client to deal with, however, it was also generally far more expensive. For example, one company had this model in place and was paying somewhere north of $23,000,000 a year for the 'subscription' apart from support. This mostly eliminated the problem the customer had, which was unexpected, choppy expenses for software. Though this amount is certainly a large sum of money, it is not unusual at the large enterprise level, and greatly simplifies the offer while allowing everyone on the implementation side of the house to get on with project.

Without this approach, the tendency of everybody is to develop their own solution, even when a perfectly suitable solution already exists on the price list. Sometimes this is because the client is unaware of the existence of these solutions. Other times, it is simply a matter of the developers wanting to develop something so they can put it on their resume. But mostly, it is because it is very difficult to have to go back to the CFO and whoever else controls the purse strings and ask for more money to purchase software. The reaction is always the same, negative.

Summary and 10 Lessons Learned

Your sales team wants to succeed and when they do, the business will succeed. But many businesses make buying from them far more difficult than it needs to be. The business that can develop a better offer, including all of the elements of the offer, will win the highly competitive game of sales. To get sales to buy-in:

- **Have the Right Tool for The Job.** First, equip them with all the tools they need to be productive.
- **Train then Train Some More.** Then train them to use those tools in a systematic way. Never forget that most training has no effect. You need to make sure the sales team begins to use the tool in their day-to-day job.
- **Be Your Own Best Customer.** If possible, try to buy something from yourself, you'll see where the rough spots are and can work on smoothing them out.
- **Constantly Improve.** Always be improving your offer. Simplify it, repackage it, repurpose it. Remember about the rule of 70[114].
- **Know that Customers Like OPEX.** Look for ways to convert one-time sales to long-term subscriptions models. It will help your customer smooth out cash flow, and often allows you to convert a purchase from CAPEX to OPEX[115].

---

[114] If you improve your offer by 1% each day, it will take 70 days to make it 2 times better. If you improved it by 2% each day, it will take 35 days.

[115] CAPEX means Capital Expenditure while OPEX means operating expenditure. CAPEX typical means very large upfront investment and that usually means getting approval from the board who will evaluate the expenditure against other CAPEX projects. OPEX normally means monthly payments, and thus, a lower outlay of cash to get going. This is normally an easier sell to a company, though the long-term cost may actually be higher. In this author's experience, the lower the initial price point and the easier the transaction, the faster you can get to the sale. If anything, the seller will often slow things down to 'qualify' the customer.

- **Be Easy to Buy From.** For very large enterprise class sales, customers make selling to them very difficult[116] which can easily disrupt your sales process.
- **Don't Let Customers Disrupt Your Sales Methodology[117] But Be Flexible.** You can push back on these onerous customer requirements. In fact, most customers expect and often benefit from you proposing a different way, whether that is at the solution level or at the sales process level.
- **Engineer Proposal Generation.** You should periodically look at what it takes to prepare an actual proposal for your product or service. Learn how to leverage tools such as Pandadoc™[118] to quickly put together very complex proposals using information from past proposals.
- **Productize Your Offer Where Possible.** If you can productize some element of your offering, it is often the easiest way to get to the first 'yes' with the customer. Don't let productization keep you from offering higher value services when needed and appropriate.

---

[116] This takes many forms and has many motivations. But from this author's experience, the larger the enterprise, the harder they will make it for a smaller enterprise to sell to them. The U.S. government is the leader of the pack of making selling to them very difficult.

[117] If you have gone through the trouble of productizing your offering, your most efficient sales process is when a customer uses it as designed. Every customer must be evaluated from a value stand point when deciding whether you want to play by their rules or walk away.

[118] Pandadoc is a CPQ or Configuration, Price and Quote tool. It is one of many on the market these days.

- **Get Sales the Right Help.** Look for ways to offload some of the administrative [119]work your sales team has perform to get a proposal out the door.

[119] When I first started with SAP Middle East, we prepared all of our own proposal responses. By we, I mean the consulting and sales team. Many of these documents were from 200 to 1000 pages or more in length, depending on the complexity of the customer's requirements. As the professional services manager, I could see then that we needed a proposal team to work these. That's exactly what we built when I returned to the Middle East (Dubai Office) in 2008. It quickly grew to about 12 to 15 people. It did not completely eliminate the work required of consultants and presales people, but now, we had a dedicated team that could develop templates, and manage our responses as the major project they were and continue to be. But just writing a response is only one part of what makes this proposal response complicated. The other is meeting all of the requirements in the RFP. This usually includes such things as providing your business' documents, i.e., corporate formation documents, proof your up-to-date with tax payments (Zakat in the Middle East), posting bank guarantees and surety bonds, and profiles of everybody on the team, including corporate officers. All of these and the many others I've seen come in require a lot of leg work. Best to take it off the shoulders of the sales team.

Summary

If you got this far, congratulations, you've covered a lot of territory. You learned how to establish and set revenue targets and work backwards all the way to how many leads you will need to generate. You've learned of the importance of having the capacity to capture the demand that marketing is generating and which sales is closing.

I've shown you the techniques for designing and managing a sales-pipeline. You've seen how to get to the magic numbers that lead you to that most important of metrics - profit.

Now you know how to figure out the amount of money you need to plan on spending for lead generation, because you most likely will want to do paid advertising to hit those revenue and growth and goals. I hope you appreciate how important this is to get right across every platform you're spending ad dollars on.

You should have come to understand that you've got to sell your plan, to yourself, to your team, and to your customers. If they don't buy off on it, no robot is going to do it for you.

I hope you've gained an appreciation for some of the challenges of managing the explosion of data and managing the flow of the information across heterogeneous systems. As they say, data is the new gold, but you have to understand how to turn that data into valuable insight and then know what action to take when you have the data in hand to turn it into gold.

You've also gained an appreciation for the rapid growth in the number of different kinds of marketing and sales systems that have come on the market. With over 8,000 systems and counting, it's a real challenge to knit together the systems you need to meet your goals. But it can be done, however, it takes time and expertise to make the right choices.

I hope you have gained an appreciation for the need to budget sufficient money to not just drive advertising, but to acquire and maintain all of the systems your business needs. It's a system engineering challenge, but more importantly, it is a business strategy problem.

Finally, you've seen what it takes to get a sales team to perform the magic of turning leads into sales. Most importantly, you've come to understand that if there is one lever above all you need to turn, it is the lever of improving your offer. The better your offer, the easier sales will be.

I hope you found this book useful and enjoyable. Within it is the knowledge and wisdom I've gained over the years from working with a huge number and variety of clients.

Index

Bibliography

Amsden, Robert, Howard Butler, and Davida Amsden. 1998. *SPC Simplified: Practical Steps to Quality.* Vol. Second Edition. Portland, Oregon: Productivity, Inc.

Brealy, Richard A. 2007. *Fundamentals of Corporate Finance.* New York, New York: McGraw-Hill Irwin.

Cascio, Wayne F. 1986. *Managing Human Resources: Productivity, Quality of Work Life, Profits.* Edited by John R. Meyer, Michael Elia and Jo Satloff. McGraw-Hill.

Gitomer, Jeffrey. 2006. *Little Black Book of Connections 6.5 Assets for Networking Your Way to RICH Relationships.* Austin, Texas: Bard Press.

—. 2007. *Little Green Book of Getting Your Way - How to Speak, Write, Present, persuade, Influence, and Sell Your Point of View To Others.* Upper Saddle River, New Jersey: FT Press.

Goldratt, Eliyahu. 1994. *It's Not Luck.* Great Barrington, Massachusettes: The North River Press.

Hana, Mack. 2004, 1999, 1995, 1990, 1985, 1973, 1970. *Consultative Selling.* Vol. Seventh Edition. New York, New York: American Management Association.

Hill, Napoleon. 1937. *Think & Grow Rich.* New York, New York: Fall River Press.

Hoffeld, David. 2016. *The Science of $elling - Proven Strategies to Make Your Pitch, Influencee Decisions, and Close the Deal.* New York, New York: Penguin Random House LLC.

Osterwalder, Alex, Yves Pigneur, Greg Bernarda, and Alan Smith. n.d. *Value Proposition Design.* Hoboken, New Jersey: John Wiley & Sons, Inc.

Osterwalder, Alexander, and Yves Pigneur. 2010. *Business Model Generation.* Hoboken, New Jersey: John Wiley & Sons, Inc.,.

Project Management Institute. 2000. *A Guide to the Project Management Body of Knowledge.* Newton Square, Pennsylvannie: Project Management Institute, Inc,.

Reilly, Tom. 2010. *Crush Price Objections Sales Tactices for Holding Your Ground and Protecting Your Profit.* The McGraw-Hill Companies.

Riefstahl, Robert. 2011. *Demonstrating to Win! The Indispensible Guide for Demonstrating Complex Products.* Colorado Springs, Colorado: Demonstrating to WIN!, LLC.

Roberge, Mark. 2015. *The Sales Acceleration Formula - Using Data, Technology, and Inbound Selling to go from $0 to $100 Million.* Hoboken, New Jersey: John Wiley & Sons, Inc.

Roetzer, Paul. 2014. *The marketing performance blueprint.* Hoboken, New Jersey:

John Wiley & Sons, Inc.

SAP AG. 1999. "TAMM 10 Materials Management Foundation."

Strauss, Neil. 2005. *The Game: Penetrating The Secret Society of Pickup Artists.* New York, New York: ReganBooks.

Tyler, Marylou, and Jeremey Donovan. 2016. *Predictable Prospecting - How to Radically Increase Your B2B Sales Pipeline.* New York, New York: McGraw Hill Education.

Weiss, Alan. 2021. *The Consulting Bible.* Vol. Second Edition. Hoboken, New Jersey: John Wiley & Sons, Inc.,.

Whitten, Jeffrey L., and Lonnie D. Bentley. 1998. *Systems Analysis and Design Methods.* Vol. Fourth Edition. Irwin McGraw-Hill.

Made in the USA
Monee, IL
04 November 2026

67380876R00157

biological ways. This is because it is designed to acclimatize in a wide range of humidity and temperature. For example, when traveling to high altitudes, the human body adjusts to ensure a sufficient oxygen supply to the body cells. The human body is adaptable and evolving through the processes of nature. The human body is continually responding to the external and internal stresses in best physiological adaptations, overcoming air and water pollution extremes, bacterial and viral infections, overcrowding, and dietary imbalances. This ability of the human body to adapt to the changing environment has made us survive in most regions of the world. We can live successfully in harsh deserts, humid tropical forests, and even densely populated cities that might have a considerable population. Contrary to humans, most other plant and animal species are restricted to only one or few environments due to their more limited adaptability.

The Earth has millions of species of life forms. The one thing we do know is... DNA is in all of them. I believe this understandable and consistent throughout the Universe. Why? Because it works. While complex, it is also simple. It gets the job done and continually reinvents itself. Not only is evolution a long-proven methodology, but mutation also does, and often, occur.

Fair-skinned humans evolved in northern climates, while dark-skinned humans developed solar protection in high heat environments; reflecting the ADAPTABILITY patterns. Land-based species have legs and water-based species have fins or flippers. If we were to look for the MOST adapted life form on Earth, I would submit waterfowl! They can walk on land, swim in, and under, the water, and fly in the air. Many are migratory. We are enjoying cool summers in the north and warmth of southernmost climates in winter. Could there be a planet of "Howard the Duck's?" Unlikely. Their wings, while wonderfully functional, are missing 'technological digit manipulation.' Since their 'life path' is adequate to THEIR EXISTENCE, DNA does seem to have some evolutionary 'stop-gaps.' Moreover, I believe we miss the mark by using the term 'alien' life. I believe it is JUST life. The other planets seem to be formed by the same physics we experienced in our solar system. A life given a suitable environment will thrive. While DNA's creational origin is still a mystery, we can simply, for now, accept it EXISTS!

Many attributes of human beings are the result of their adaptability and survival, known by the insights of anthropological studies. It is a known fact that as evolution takes place, living beings adapt to the changing environment and exist in the most updated design. The inherited traits in humans are the characteristics that our ancestors conferred to us. The basics of anthropology, the culture and linguistics, also help understand the behavior and adaptation of human beings. Culture is a shared aspect based on five characteristics. It is shared, learned, uses symbols, is dynamic, and is integrated. Each culture is integrated with these fundamental elements. Culture is, however, not biology as we do not inherit culture from our ancestors. Cultural anthropologists study and explore the sociocultural systems associated with a human being around the world. We are a chimera of culture, language, traits, norms, intellect, inherited characteristics; evolving, adapting and surviving alone since 4-billion years. Human genetics is rapidly evolving, enabling adaptation while responding to increased environmental stressors. The most predominant human attributes interchangeably referred to as human qualities, including integrity, courage, honesty, whole-heartedness, and self-awareness. These qualities define human nature on a universal plane and ingrained in different cultures. Human traits that ensure survival include reproduction, physical strength, keen eyesight, tendency to engage in social behaviors, and resistance to fight the disease on the biological plane. With evolution, these biological and cultural traits have advanced, which has also been accelerated by technology, innovation, increased life expectancy, and better medicines. There is still a need for scientific exploration of the power to live that encompasses different natural disasters parameters and other adversities[31]. These speculations are necessary to investigate the qualities needed by humans for survival in the evolutionary process. Studying human contact experiences with beings from other planets requires rationale thinking and insights on what evolutionary aspects can lead to these contacts? What would be the fate of humans after they have successfully contacted extraterrestrial beings? Would this contact be harmful or beneficial?

Explorers of extraterrestrial life have provided substantial grounds for rationale thinkers to contemplate the importance of extraterrestrial contacts. Research in this context is mostly anecdotal; however, it can be compiled based on similarities in these contact events and experiences. One clue to how we prepare for formal contact begins with a computer analysis of 'messages' both past and present that allude to non-terran contact.

---

[31] Sugiura, M., Sato, S., Nouchi, R., Honda, A., Abe, T., Muramoto, T., & Imamura, F. (2015). Eight personal characteristics associated with the power to live with disasters as indicated by survivors of the 2011 Great East Japan Earthquake disaster. *PLoS one*, *10*(7), e0130349.

For example, Sumerian ad Mesopotamia was foundational in the Middle Eastern' Cradle of Civilization.' In Sumerian, the origin of cuneiform writing emerged. From a hunter/gatherer culture and almost overnight society of writing, math, engineering and governance seemed to manifest itself instantly. Yet, as mysterious as it looks, chronologically, the Sumerians had an answer themselves. The Annunaki. Their 'Gods who from the Sky came.' While current archeology insists this is mythology, the lifelong Sumerian scholar, Zacharia Sitchin, much published himself, indicated this was more history than a myth. This 'instant evolution' was experienced around the world. Almost identical construction techniques followed by peoples who never had any verifiable contact! 'Gods' with similar names as found in South and Central America. Feather serpent gods, who were only separated by linguistic differences between cultures, performed equal feats of wisdom and imparted knowledge.

Some scoff at the incidents, stories and sightings as hysteria or misidentification of other phenomena. The problem is - they are TOO PREVALENT to overlook. So, it is logical to believe that SOMETHING is amiss. I think 'they' do not indicate WILLING INTERACTION. This can be easily understood as little is being done, as a society, to accept their presence and initiatives to make it so. SETI is relatively passive. And more importantly, I do not know any government program to PREPARE for such an eventuality. I would suggest that any course of analysis and study must include creating 'Close Encounters of the Third Kind'-type scenario. This must NOT be left to a 'someday' goal. Let's suppose we had a prepared site at the top of Devil's Tower or other remote location.

From there, the broadcast of signals and language variations could be transmitted NIGHTLY in every technical form we could conceive. Audio, Visual, Telemetry, Binary, Laser and other communicative forms would be sequentially sending 'We're Ready to Meet' messages. Some of these methods could be made mobile and sent to places where CURRENT 'flaps' occur. The Message- GO HERE! Using the aforementioned site prep location. This is a significant undertaking. Undoubtedly, considerable government funding would be required, but that is what it's for! We have diplomatic outreach all over the world. Some off-world, especially if the contact could lead us into a new technological era, is more than worth the effort. We have Nothing to Lose and Everything to Gain.

It is essential to enlist the similarities in the contact experiences that have occurred so far in history. The origin of UFO sightings can be traced back to the civilian pilot, Kenneth Arnold's flight, in 1947, stating sighting of "nine blue, glowing objects flying fast—at an estimated 1,700 m.p.h.—in a 'V' formation." Later in 1951, three science professors from Texas Tech saw a semi-circle of lights flying above them at a very high speed. In 1957, dozens of citizens of Texas reported strange lights or rocket that interfered with their vehicles. These were reported as an electrical storm that caused mechanical malfunctions. These incidents took place in Texas. Most of the UFO sightings are reported to occur in Western suburbs. Reports of triangular UFOs hailed from across the US in the 1950s, 60s, and 70s. Mysterious flying triangular objects were reported during the 1960s, over Georgia, Connecticut, Texas, and Pennsylvania. Other similar triangular objects were also reported from Madrid, London, and Czechoslovakia. Two National Guard pilots reported 'triangular-shaped object, 50 feet in diameter' in 1969, over San Juan, Puerto Rico. The officials attributed most of these incidents as weather balloons, atmospheric conditions, and other outer space objects; however, some of the incidents had remained unexplained[32].

Most science areas, such as semi-quantum kinetics, electrohydrodynamic, graviton-electrics, and electrogravitics are classified as the black areas of experimental and theoretical physics. This is because these sciences tend to define the theories behind alien means of propulsion. For example, exploring the semi-quantum kinetics, electrohydrodynamic, gravito-

---

[32] Daugherty, G., & Sullivan, M., (2020). Huge, Hovering and Silent: The Mystery of 'Black Triangle' UFOs. https://www.history.com/news/black-triangle-ufos-facts

electrics, and electrogravitics of alien means of transport would answer controversial questions such as do aliens need any kind of fuel for traveling? How do they travel such long distances? Why do they always seem to travel faster than human-made machines? Information gathered from the incidences of extraterrestrial contacts has let the human race rationally think on this topic and shape different questions based on this information. Most UFO sighting incidents led to the foundation that they travel at speeds higher than a human-made machine. However, modern-day physics fails to explain the science behind these speculations. Many facts are still unexplainable such as disappearing of UFO immediately after they are sighted, flying objects in unusual shapes than those that can be made from known mechanistic and thermodynamic principles[33].

Amateur research, entitled ufology, is sparked after accumulating factual information on UFOs and alien encounters. Moreover, these researches are

---

[33] Bhattacharjee, D., Harikant, A., & Roy, S. S. (2020). A Scientific Study of the Unidentified Flying Objects in accordance with Anti-Gravity.

accompanied by government investigations. Part of the scientific community has greeted UFO research and exploration with skepticism calling believers of UFO and extraterrestrial beings as abnormal or irrational[34]. The information and data collected so far in UFOs and extraterrestrial life are not enough to conclude the presence or absence of extraterrestrial beings around us. Despite substantial evidence though anecdotal, contradictors and naysayers require further explanations to believe in UFOs, extraterrestrial being. Americans have reported UFO sightings and contact experiences for more than half a century in an up-close way. Despite several reports, the community of contradictors fail to accept the concept of UFOs. Some incidences encouraged the critics presented by this community such as Kenneth Arnold's move to monetize his report when he sold the story to "Fate Magazine." Similar observations have increased the scientific community's

---

[34] Eghigian, G. (2017). Making UFOs make sense: Ufology, science, and the history of their mutual mistrust. *Public Understanding of Science*, *26*(5), 612-626.

skepticism in accepting extraterrestrial research as a part of scientific exploration. These critics might be a significant challenge in our journey towards UFO exploration and contact.

Chapter 3

Variations on Species is Consistent

The link between UFOs and the existence of alien life has long been discredited as pseudoscience. However, the increasing number of reports on UFOs, their witnesses and government disclosures have recently surfaced, which has challenged the notion that 'UFOs and alien existence belongs to pseudoscience.' Due to the gap between urban legend-like stories and verified Unidentified Aerial Phenomena (UAP) sightings, it is difficult to believe in the possibility of alien visitors. Although, the controversial debate on the topic has increased the number of questions and judgments but also enhanced the level of curiosity among intelligent life hunters and explorers of extraterrestrial beings. Most of the sightings are unverifiable, where individuals claim that they have encountered UFOs and aliens. The blind ends to the UFO-lined debates are further advanced by the number of hoax-like cases in which the witnesses are often motivated by the financial gains and publicity from their alleged experiences. Nevertheless, certain cases also exist in which groups of adults and young children have witnessed strange entities emerging from UFOs, making it difficult to attribute such an anomaly to human psychology.[35] Based on these rare incidences, this section of the book is focused on identifying and putting forth the species of UFO occupants that share common characteristics.

Information gathered from various witnesses indicates that UFOs generally appear in four main shapes; the cylindrical, the saucer or the disc, the sphere, and the enormous triangle. Also, they shoot out beams of light, illuminating a terrain beneath; exhibiting brilliant luminosity. The beam is sometimes truncated and often

---

[35] Han, G. "Aliens" and UFOs.

has a pulsation of a full spectrum of colors. Sometimes they appear in cloud-like lenticular forms and are frequently surrounded by vapors. UFOs witnessed mostly have portholes, domes and have a metallic-looking, light-colored, shiny surface. It has been observed that UFOs have an unusual effect upon their immediate surroundings. In presence of UFOs, animals may behave strangely and often panic. Usual electrical malfunctions are caused by heavy electromagnetic effects. Car engines may stop running as well as the headlights and radios, as soon as they are in close proximity to UFOs. All systems startup on their own upon the departure of the UFO. There may be signs of radiation effects on animals, plants, and humans in their immediate areas. They may also feel temperature fluctuations such as feeling hot or cold.

There are variations and similarities in the behavioral patterns of UFOs. Most often, they do a prolonged, motionless hovering, which later propagates into extreme acceleration, straight upwards. Most UFOs show non-linear, erratic, zigzagging, non-smooth, and darting motion. Sometimes the vehicles stand upright in flight, or flip end over end, or fall like leaves. They are designed and able to make sudden right-angle turns at great velocities. They exhibit super-sonic speed without any sonic boom. Sometimes there is just a high frequency with low volume humming sound, and rarely any sound made during their movement. Most of the UFOs witnessed leave certain after-effects on the ground. Most of the impressions are in geometrical shape or design, having a symmetry. Vegetation and soil in that area is affected, becomes dehydrated, and may not absorb water for long term. Moreover, affected vegetation in UFO exposed areas may not propagate or regrow for a considerably long time.

Extraterrestrial life explorers have also analyzed the features and characteristics of aliens that resemble human traits. According to Aggarwal (2017), aliens are shaped by the same mechanisms and processes that have shaped humans, such as the process of natural selection that has enabled the human race to survive and thrive on planet earth[36]. According to a study published by the University of Oxford in the International Journal of Astrobiology, evolutionary theory can be used to better understand the behavior of aliens and to support alien predictions. It also supports the well-accepted argument that extraterrestrial beings also undergo a natural selection process similar to humans, enabling them to evolve stronger and fitter over time.

Sam Levin, a researcher in astrobiology in the University of Oxford, states, "... Making predictions about aliens is hard. We only have one example of life—life on Earth—to extrapolate from. Past approaches in the field of astrobiology have been largely mechanistic, taking what we see on Earth, and what we know about chemistry, geology, and physics to make predictions about aliens." Reports of Abduction are usually scorned and scoffed at, yet, curiously, they take many of the same

---

[36] Aggarwal, M., (2017). Aliens may have a lot in common with humans: Study. https://www.livemint.com/Science/YTVVSS0DCl993dpGypccPM/Aliens-may-have-a-lot-in-common-with-humans-Study.html

form. 1. The person{s} are remotely located. 2. All indicate some form of 'Lost Time,' that period between contact and when they actually realize the situation is over. 3. A medical procedure of some kind is performed, many times in reproductive areas. 4. The conduct of the 'aliens' seems totally immersed in their mission and little to do with the emotion or pain of the subjects. This may be a vital clue as to the species' origin. I am of the belief the Grays are a previous cephalopod nature. The same as cuttlefish, squid, or octopi. This would be a reasonable assumption by our own earthly studies. There is no research I am aware of that cephalopods experience emotion. A squid does not size up its next meal or wonder about the prey's family. It considers, is it close enough? Is it a recognizable food form? Then they act. These are the same traits of Grays: Obviously Intelligent. Technological advances we have yet to understand. However, they are seeking something! I feel that their curiosity is a technological one, not emotional. Indeed, it may be the emotional construct of humans that fascinates them. It has to be something uniquely human. Their biological knowledge may be of their own world and existence, but we're not odd in similarity. We both respire. We both are animated flesh. We are both 'living' and some form of ingested nutrition is likewise in place. But this is true of all our wildlife too! Therefore, I believe they just want to know "what makes us tick." A scientific endeavor where we have the obvious subject matter.

Predictions on the biological make-up complex aliens have also been made responding to queries like "what aliens might look like?" However, still it is unsure if aliens would walk on two legs or have big green eyes, as shown in science fiction movies. These insights are supported with evolutionary theory, offering a unique tool to make us understand what aliens are like. On the basis of evolutionary theory, strong predictions can be made by considering the major transitions that aliens have undergone. It can be said that there might be a level of predictability to evolution that would make aliens resemble us. There are potentially thousands of planets that are habitable in our galaxy alone. We have already taken the step to answer whether we are alone or not. If we are not, who are our neighbors, and what do they look like?

Most of the insights on alien existence has been taken from information gathered from the incidences of alien abduction. Also called UFO abduction, or alien abduction syndrome is recognized as a personally held belief in which the abductee explains a subjectively real experience of their secret kidnapping by the non-terrain entities. This concept has also been perceived by most mental health professionals and scientists as conditions such as psychopathology, sleep paralysis, deception, and suggestibility. Common claims that have been made involve forced medical examinations[37]. Other abductees claim that they got warnings of the dangers of nuclear weapons and environmental abuse. "Abduction" phenomenon is recognized as one of the most enduring mysteries for the Alien-Human interface. Clearly, this has sociological importance as we try to understand these beings' motivations and, therefore, their socio-cultural constructs. First, their actions cannot be put, realistically,

---

[37] Dagnall, N., Drinkwater, K., & Parker, A. (2011). Alien visitation, extra-terrestrial life, and paranormal beliefs. *Journal of Scientific Exploration*, 25(4), 699.

into human terms such as 'Abduction.' abduction is a human moral violation. Yet, we have no problem 'abducting' mice and rabbits for medical experimentation. Humans have developed a series of moral imperatives we call 'Law.' They may have similar 'rules' for social interaction, but we have to view it from their perspective to understand their actions.

Secondly, there does not seem, in most cases, any direct malevolence in their procedures. That is, for any person{s} to report what happened, they had to be returned! Every year thousands of people go 'missing' from Planet Earth. Missing person reports are voluminous. Yet, many have earthly resolution. Runaways; abuse by spouses; tired of current circumstances, in general, and even serial killers are involved. Can some ultra-terrestrial entities take someone permanently? Of course, such is possible if you accept the premise that it occurs at all. Nevertheless, while we have found mutilated cows, sheep, horses, and the like, no dissected human has ever been noted in a UFO related incident! This would, at least, imply that some level of distinction is made in the case of human examinations.

Thirdly, reported scenarios are almost identical. Lonely area or road; taken into a 'craft'; an examination room and table; procedures centered around sampling; regaining consciousness with little or no clue as to what actually occurred. Based on sample taking stories, there seems to be a genetic component sought. There can be many reasons, however; I have my own view: Since other animals are taken and dissected, and we are not, they recognize >our< level of concern for non-human species of our world and our higher concern for humanity.

Next, they appear to be seeking the difference in our biostructure compared to these other samples. This would indicate that they understand we are the dominant species on this world and that bears comparative inspection, to our own life forms, they encounter here. Ultimately, I think they have some genetic issue they are trying to resolve. It could be anything as complex from hybridizing some life form for populating a sterile planet in their sphere, OR it may be as simple as a personal reproductive degeneration or disease for which they have no solution in their own environment. Moreover, they do so with impunity. Some might say, "Hey, come down, introduce yourself, we can have a cross-exchange of medical advice and methods for our mutual benefit. We may not have that kind of, well, respect, to which they are willing, at this time, to express! If they have been coming here as long as some surmise: Wars, Hatred between cultures and races, uncooperative government to government assistance, on a global scale, may not fit their perspective of a truly advanced species. Observations, radio and television signals, including 'news,' and other monitored communications, may not put our 'best foot forward' to non-terran visitors! By our own hands, we may be getting dissed by the galactic community.

In the digital age, compactness is ever sought. Smaller and smaller devices holding more processing potential allows more to be done in an economy of space. This is not the natural world. Mankind is the dominant intellect on Earth. However, there are closely equal animals in the world that very intelligent. Elephants porpoises, simians and even octopi. The difference between man's brain, and that of the elephant, for example is certainly smaller. But our biometry allows it to be exercised and asserted more efficiently. Two significant differences: 1. Our ability to conceptualize and 2. the opposed thumb. The articulation between the thumb and fingers is masterful of technology. We can turn knobs, flip switches and hold tools. This directly exercises the conceptualization of what we can do! Elephants while intelligent have four legs of ponderous and stumpy appendages.

Their 'evolutionary imperative' is limited to travel for food which they consume vast amounts of vegetation. Yet, this is sufficient for their ambition and lifestyle. On the other hand, human's bipedal locomotion gave us migration and efficient travel for hunting and gathering, hands to create shelter and clothing to face the elements of whatever climate was to be inhabited. What I call 'corporeal engagement' dictates giving evolution a hand in expanding an intelligent being's intellect. As noted earlier, it is not an accident that our brain capacity has grown significantly along with the development of knowledge and nutritional advancement. Our ability to communicate that knowledge shares the growth with others of our species. These are the same qualities we must search for in SETI effort. The Universe has dictated who is to be a predominant species in a given environment. Moreover, any interspecies engagement will depend on bio-symmetry as a productive encounter. We seek out beings that can converse and have equal, or undoubtedly superior, intellectual commodities. While interesting and curious, it is unlikely that an intelligent sponge, from some watery distant world, wouldn't have much in common for us to share.

Ray Bradbury, sci-fi writer extraordinaire, probably described it best in 'The Martian Chronicles." When a Martian was queried by the Earth's expedition leader on the Meaning of Life, it replied life is its own answer. It is nature's way of experiencing the totality of being." The Universe has decided that DNA is the agent provocateur. The biological macromolecule that ensures the transfer of traits from one generation to another, sustaining the human race on planet earth. But considering the presence of the same molecule leading to the existence of extraterrestrial life is complex as our knowledge does not encompass the biological aspects of aliens or even their presence. Maybe they are made up of the same carbon and nitrogen like you and me, and we are simply unable to see them or maybe not. If such life exists, it would sustain in the 'shadow biosphere.' This might not be considered as a ghost realm, but unidentified and undiscovered creatures probably with a distinct biochemistry. Since they are outside of our comprehension, we can't study or even notice them; a major restriction for us to know and learn more about these beings. If that shadow sphere really exists, why haven't we found it yet? We have the tendency to study the microscopic world, but in limited range, as only a small percentage of microbes can be cultured in a lab. This reflects the fact that there might indeed be several other life forms that we have not spotted yet. We can only detect life if it contains DNA - otherwise we are unable to explore other life forms. Panspermia, the theory that life travels by meteor or comet between worlds, is a reasonable postulate. We know what Earth like conditions can achieve with an existential supply of DNA, therefore, I believe it is fair to say other planetary conditions will utilize it appropriately. Could there be a planet of Intelligent Machines? Perhaps. However, there would be a biological creator to assemble them, initially, from raw materials. Thus, we have the same concerns over Artificial Intelligence here in our own world. I agree it is extremely important that AI safeguards be implemented as it advances. Intelligent 'devices,' without empathy for its creators, could be very problematic. This is why the challenge of finding Intelligent beings and contacting them is so exciting and wondrous. What challenges have they faced and overcome? In technology, culture, medicine, sciences, co-existence. What solutions for the human condition may we benefit from their going down those roads' eons ago?

Genuine stories of alien abduction go back to the

1950s, the case of Antonio Villas Boas. Later in 1961, the case of Betty and Barney Hill abduction received widespread publicity. The Hill incident was recognized as the prototypical abduction, which was the first leading to the information that beings were explicitly identified as extraterrestrial in origin and these beings later became widely known as Greys - the most common types of extraterrestrial in the reported abduction cases. To my knowledge, we have not even got into the initial phases of extraterrestrial contact and require strong justifications to address the debates and controversies associated with the features, structures, shape, and biological aspects of aliens - if they exist - as our grand query to resolve. Most pieces of information reflecting the type and nature of intelligent lives poorly represent the traits and features of aliens, which leads us to blind ends, failing to answer our questions. This can be attributed to various corollaries that we might think of while considering the presence and characteristics of extraterrestrial life forms. From my observations a number of corollaries seems probable: 1. Digital intelligence can be extremely compact; Bio-intelligence requires some volume. 2. We have numerous examples, on Earth, of 'intellectual distribution.' 3. Intelligence alone is not enough. 4. Locomotion is required to implement intelligence.

Among the so-called accepted and genuine cases of UFO contacts, Roswell UFO incident is one of the popular ones, referring to the speculations and rumors regarding a balloon crash at a ranch near Roswell, New Mexico. Interest in this incidence increased after the ufologists started promoting the conspiracy theories claiming that extraterrestrial occupant was recovered by the military as one or more alien spacecraft crash-landed, that was covered by military personnel's conventional weather balloons. The Roswell incident has been recognized as "the world's most famous, most exhaustively investigated, and most thoroughly debunked UFO claim." Claims have also been made for the presence of alien bodies in Area 51 at Groom Lake in the United States. The most famous claim holds that this area hosts alien aircrafts as well as the bodies of its pilots, after their crash at Roswell[38]. A man named Robert Lazar has also claimed in 1989 that he had worked on alien technology inside Area 51. He also claimed to have seen the images of aliens, and the facility is being used to examine the UFOs. The association of Area 51 may have served the intelligence agencies as a useful distraction. To our knowledge, alien related myths and claims still require strong evidence to make people believe in their existence. The comparison of intelligent life with human life may address the fallbacks ufologists are experiencing in their search. The questions associated with the color, shape, living attributes, and behavior of extraterrestrial life still need substantial investigation and rational thinking capacities. We might not be able to contact them but our quest to identify their existence might unleash realities underpinned in human and other life forms on Earth.

---

[38] BBC, (2019). What is Area 51 and what goes on there? https://www.bbc.com/news/world-us-canada-49568127

Chapter 4
The Gray Race

The evolution of human beings is researched with two approaches, top-down and bottom-up. Darwin's theory of evolution explains the bottom-up approach, where we trace molecular evolution from simple to complex, to explain life's origins. Extensive lab experiments have been conducted to prove that it is possible to create amino acids with simple precursors in the presence of energy. The other approach is to try and trace evolution backward from extant life forms to their earliest microbial ancestors. By identifying their ancestors, we can assess the conditions they were born under, and what factors in their environment may have led to their evolution into the beings they are today.

These two approaches of researching human evolution are complementary and converge at a certain point, proving their authenticity and reliability. It is clear that the emergence of life on Earth can be linked back to the environmental conditions present on the planet over three and a half billion years ago. Understanding the conditions that this life arose under leads us to the belief that this evolution has occurred on other planets in and outside our solar system as well.

The conditions of prebiotic Earth billions of years ago were heavily influenced by factors like the sun's luminosity, the frequent inrush of meteors, comets, and apocalyptic impacts, and the tidal force influenced by close proximity to a large moon. At that same time, similar conditions must have existed on Mars as it is assumed that its massive volcanoes were active then, and its surface temperature and pressure was enough to sustain and support the presence of water. Therefore, research on extraterrestrial lifeforms and beings is extremely dependent on the exploration missions to Mars and discovery of life there.

Similarly, the advancement of observational astronomy has enabled us to discover a multitude of planets outside our solar system that seem suitable to support life. As this technology advances, we may soon be able to detect traces of microbial life on these distant planets. Our planet has supported microbial life for over seventy-five percent of its existence, that's over three and a half billion years, but only within a hundred years, we developed technology that enables electromagnetic communication. It's safe to assume, based on how technology is advancing exponentially, that soon, our search for microbial signatures and extraterrestrial life will be successful.

Until we can find microbial life, we can only employ the top-down approach to our research on alien evolution based on the recorded incidents of the few lifeforms that have visited us. The 'gray race' has been the most commonly described species of alien lifeform that has been sighted on Earth. Popularized by pop culture and cinema, these aliens are described as being small humanoids with large heads and massive eyes. They are classified as the gray race due to the greenish-gray pigment of their skin. They possess reduced forms of or completely lack external human organs such as ears, noses, and reproductive sex organs. Their skin is described as being smooth and hairless with small openings for a nose, mouth, and ears. Their bodies are generally slender with elongated abdomens and short, differently jointed legs. Gray aliens are also said to be a lot shorter than average humans, with little to no muscular definition and their skeletal structure visible through their skin.

There are many links and references available to the gray aliens in archaeological findings, as well as first-hand experiences of some individuals.

The most notable and popular case once again is that of the Roswell UFO crash. Locals claim they saw military personnel carrying grey humanoid bodies on gurneys at the UFO crash site. In 1992, a cameraman who worked for the US army and air force when the Roswell incident occurred released a photograph, a still from a video where an alien is seen lying on an operating table[39]. The original footage was never released, but a

[39] Express, (2017). Is this a REAL alien? Claims this is a

terrifyingly close example was recreated in 1995, showing the alien very clearly as compared to the grainy negative from the original. The presumably dead alien lying on the table perfectly fits the description of the gray species of extraterrestrial beings.

The archaeological site, Puma Punku, also has various references to grey aliens, with the most notable being the embossed inscription of an alien head in, what is known as, The Wall of Humanity. Many believe that the sculpted heads present in this wall represent different tribes, but they actually represent the different alien species that have visited Earth. Puma Punku is proof that aliens have been visiting us for thousands of years, perhaps more.

In 2017, mummified remains of what are said to be three-fingered alien bodies were unearthed by Russian archaeologists in the Nazca Tombs in Peru[40].

'genuine 1947 Roswell alien autopsy image' https://www.express.co.uk/news/weird/860020/Roswell-alien-autopsy-genuine-alien-image-Ray-Santilli

[40] Evening Standard, (2018). Three-fingered Peruvian mummies unearthed in tomb last year 'have alien anatomy'

Six bodies were found in the tombs present in the city of Nazca, Peru. These bodies are very small, the size of toddlers, with proportionally larger heads and eye sockets. Their tissue is biological with twenty-three pairs of chromosomes in their DNA, just like humans, but they aren't exactly human. Their anatomic structure suggests otherwise. An additional twenty mummified bodies were found in Nazca, holding a very strong resemblance to the gray alien bodies found at the UFO crash site in Roswell[41].

Two percent of all extraterrestrial sightings reported in the United States were of gray aliens, which is considerably higher than any other nation. In 1961, an American couple, Barney and Betty Hill claimed to be abducted by aliens. This was the first widely publicized case of alien abduction. They were abducted in a remote part of the state of New Hampshire by gray aliens

---

https://www.standard.co.uk/news/world/russian-researchers-claim-threefingered-peruvian-mummies-unearthed-in-tomb-last-year-have-different-anatomy-to-humans-a3788421.html

[41] Express, (2017). Nazca Tomb: Shock 'Roswell link' made in '20 alien bodies found in tomb' probe https://www.express.co.uk/news/weird/847189/Nazca-aliens-Nazca-tomb-Roswell-UFO

their highly advanced inventions, language, infrastructure, and art, the Sumerian civilization is the oldest known civilization. Hieroglyphs are more popularly associated with the Egyptians, however, and even some Chinese civilizations used hieroglyphs for some time.

Zechariah Sitchin, one of the few scholars able to read and interpret Sumerian clay tablets, found that the Sumerians used to worship a group of Gods known as the Annunaki, who descended from the sky and communicated various ideas, skills, and technological methodologies that came to life after their visit. Not just the Sumerians, civilizations from East Asia and South America all have references to “sky Gods” in their archaeological scripts. The fact that the Mayan calendar, which is a stroke of mathematical and astrological genius, was created during a time when all other nations of the world were busy hunting clearly shows that some kind of external influence was involved in the advanced methodologies of these civilizations. According to Sitchin, the Annunaki weren’t Gods but advanced human-like extraterrestrials that resided on a planet beyond Neptune, the 12th planet, known as Nibiru. Sumerian scripts claim that the Annunaki were gray beings with tiny mouths and large craniums, a description that matches that of general aliens, and that they communicated with people through telepathy, indicating that perhaps they have evolved beyond human comprehension and can employ means such as telepathy to communicate. Sitchin's theory of the Annunaki being extraterrestrial beings may be true because, for people in those times, any unnatural phenomenon such as beings ascending from the sky would have been considered divine and godly.

Sitchin's claims about the Annunaki might seem plausible to people who believe more in the religious or spiritual aspects. However, Sitchin's theory contradicts the beliefs of many scientists who suggest that if extraterrestrials could detect us based on street lighting, they would have done so some 140 years ago, and if they could detect us from radio waves, that would have only been possible after the first radio broadcast by BBC in

1922. Otherwise, it would have been highly unlikely for planet earth to be detected in the galaxy by life forms on other planets. There are two possibilities to both these claims; either the Annunaki were really Gods and not aliens, or since we still do not know how advanced extraterrestrials are, they must have other ways of detecting our planet apart from radio waves and light. The latter seems more plausible.

If it weren't for scholars such as Sitchin, the mysteries that ancient scripts and tablets hold about how ancient civilizations operated and how the world has evolved since then would perhaps remain a mystery for us for all eternity. This further strengthens the debate of how Exo-Anthropology requires more attention. Linguistics should be a major area of focus under Exo-Anthropology. How ancient civilizations came up with the idea of hieroglyphics to record important events and how they transitioned from hieroglyphs to scripts that did not involve pictures over time is yet another mystery, and it makes you wonder how it all started.

The most vital form of communication is language. Suppose ancient civilizations came in contact with aliens and were influenced by them. In that case, it is quite possible that studying the way they communicated could be the key to communicating with aliens, and in order to do that, we need more scholars who can help decipher ancient tablets and scripts of text. It all comes down to one thing; promoting the field of Exo-Anthropology to better understand other life forms. Different words or hand gestures mean different things in different cultures. For example, the thumbs-up gesture for us means "good job" or "good luck", but in certain countries such as Greece, China, Malaysia, and Singapore, it is just as offensive as the middle finger. It is important to keep cultural barriers in mind when setting up a system of communication with extraterrestrials. It is sometimes hard to communicate efficiently with our own fellow human beings, so how can we expect to establish effective communication with other-worldly beings?

The frequent observation of Unexplained Aerial

Phenomena (UAP) has established one common factor between us and extraterrestrials: that we are both curious beings, and just like us, they possibly wish to make contact with us. Sooner or later, we are bound to make direct contact with these extraterrestrials beings, and so, we should be prepared for it when the day comes. The wars, famine, and inhumane treatment of fellow beings around the world do not make for the best of impressions on other planets, and so establishing intention upon contact is vital to present ourselves as beneficial members of the community of larger intelligent species.

The field of research for Exo-Anthropology does not have to be finite. It can be an ongoing program to which new researchers can continually contribute. An annual EA Convention would also be an excellent idea to promote EA amongst the scientific community, and having guests and panel members from a UFO knowledgeable community can provide much insight on how to move forward.

Technology has come a long way, and while it

may take some effort, it is not impossible to devise computer software that can interpret and translate all languages known to man. This can be beneficial in the field of Exo-Anthropology, and a contact management team comprising multilingual international members can also be formulated. The technological advancement that has taken place in society in the last few decades is astounding. If things keep progressing at the same pace, who knows where we will be in the next 100 years?

Just because aliens seem to be much more technologically advanced than us is not to say that we might not have a few things that attract other life forms from other planets. Throughout history, UAPs have both shocked and fascinated the masses, and many a time, these have occurred near military bases. During the end of World War II, in 1944 or 1945, US airmen reported having seen strange orange lights flying in the sky near the French-German border, which they called "foo fighters". During the Korean War, in the early 1950s, soldiers claimed to spot "pulsing rays" being emitted from a blue-green light, which also made them sick.

EXOANTHROPOLOGY: A NEW SCIENCE FOR THE 21ST CENTURY'S FIRST CONTACT

While governments have always tried to keep information regarding UFO sightings and UAPs from the public, high-ranking US military and intelligence personnel have revealed that they have witnessed UAPs near nuclear power and weaponry sites. UFO sightings have increased near nuclear power plants ever since the H-bomb tests carried out by the United States in the Pacific Marshall Islands. In the late 1940s, "green fireballs" were spotted in the skies near atomic laboratories in Los Alamos and Sandia in New Mexico. "All of the nuclear facilities – Los Alamos, Livermore, Sandia, Savannah River – all had dramatic incidents where these unknown crafts appeared over the facilities and nobody knew where they were from or what they were doing there", says investigative journalist George Knapp who has studied the relation between UAP and nuclear power for more than 30 years. In 1967, a UFO was sighted at the Malmstrom Air Force Base in Montana, which contained nuclear-tipped ballistic missiles. The air captain, at the time, later stated that many of the missiles stopped functioning at the same time that a glowing, red object was seen hovering over the facility. Another famous UFO sighting is known as the Bentwaters-Rendlesham Forest incident. In late 1980, air traffic controllers spotted something very strange near the Royal Air Force Bentwaters in England, which was home to 25 secret nuclear weapons. A giant, basketball-shaped unidentified aircraft stopped over the base for a few seconds before disappearing in the blink of an eye.

The fact that UFO sightings have suddenly increased over the years, especially near military bases and nuclear power plants, indicates that something about this powerful technology seems to attract extraterrestrials. Perhaps the governments have tried all these years to hide this information from the public is because the frequent appearance of UFOs near such areas could pose a potential threat to humankind.

Considering the state of our technology currently, it seems that it is not as easy for us to communicate with extraterrestrials as it for them. While the Search for Extraterrestrial Intelligence (SETI) program aims to detect signals or any evidence of technological civilizations that may exist on other planets, the Messaging to Extraterrestrial Intelligence (METI) is a branch of study that creates and transmits interstellar messages towards habitable planets in an attempt to communicate with extraterrestrial civilizations. So far, however, no such signal has been detected, nor has any organized or clear message been interpreted yet.

While these programs should be continued, there is a debate between scientists on whether we should continue the efforts to contact extraterrestrial beings on other planets or cease them, for they might visit us in the future. We cannot know what intentions they might have, especially considering the fact that they seem to be attracted to our nuclear weapons.

However, we should still continue to make a vigorous effort to determine if such beings actually exist or not.

Chapter 8
Where Do They Come From?

Are we alone in the universe? That is a question frequently asked by scientists, researchers, astrologists, and the common man alike. Many space missions and other ground-based observatories have been deployed over the last few decades to detect the existence of life on other planets. One such task was assigned to the Kepler mission, which was launched by NASA on March 7, 2009, and ended on October 30, 2018. Its objective was to explore a part of the Milky Way to detect planets outside the solar system orbiting other stars as well as determine how many of these millions and billions of stars in the whole Milky Way have such planets orbiting them. During its 9.6-year run, the Kepler space telescope discovered 2600 exoplanets after observing more than half a million stars in the Milky Way. 550 of these planets are said to be possibly rocky planets, and nine of these planets orbit in their stars' habitable zone.

This shows that our solar system is not the only one in the universe. New solar systems with hundreds of new planets are discovered every year, and scientists estimate that there could be billions of solar systems in the Milky Way alone. Considering that there could be anywhere between 200 billion to 2 trillion galaxies in total in the whole universe, it is hard to imagine just how vast the universe is and how many millions and billions of other planets exist out there. One question that remains unanswered, even after the discovery of new planets and galaxies from time to time, is whether life – simple or complex – exists on these other planets.

"Where are they?" asked Italian physicist Enrico Fermi about the existence of aliens in 1950, which led to the development of what became known as the Fermi paradox. According to his theory, there are billions of other Sun-like stars in the Milky Way, all of which have a high probability of Earth-like planets orbiting them. Since our solar system is not the only one in the universe and is only about 4.5 billion years old, others are likely to have been around for much longer, meaning intelligent life on them, if it exists, must have developed long ago. What makes his theory so contradictory is that even with the high probability that there are millions and billions of other Earth-like planets in the universe, and many of them must contain life, we should have been visited by these extraterrestrial beings already, or in the very least, established contact with them, but so far, there is no solid evidence to suggest that such a visitation has occurred, not has contact been established with any extraterrestrial life forms.

So, what are the chances that intelligent life does exist on some of these planets? What if we are not alone in the universe? If so, where do these extraterrestrial life forms come from, and what conditions do they live in?

Earth makes an excellent lab for understanding extraterrestrial environments and how they affect the various life forms or "configurations". Configurations can best be described as life forms. This term has been coined because the more we discover about living creatures, the more diverse they become. When talking about life on other planets, we primarily have anecdotal information. However, life on our own planet is so diverse that it makes you wonder what else could be out there in the universe. Take reptiles, for example. In the life form laboratory that we call Earth, reptiles are found all over the world. Interestingly, they have local adaptions just the same as they have species diversion. Reptiles can be scaly, smooth or feathered, depending on the evolutionary path or environment they live in.

Much of the time, the population loses sight of common marvels. As the only planet we are absolutely sure has life, the great abundance and variety of different life forms often escape constant attention. We have every possible environment where living creatures have evolved and adapted themselves to changes in the environment and survived over hundreds and thousands of years. Animals, for example, depend on their physical characteristics to survive in their respective environments. In the Arctic, animals have special adaptations which allow them to combat extreme conditions. Polar bears have white fur, which enables them to blend in with the snow and hide from prey, thick layers of fat and fur protect against the cold, and their large feet evenly distribute their weight and help to grip better on the icy surfaces. Penguins, seals, and whales also have thick layers of fat called "blubber", which acts as an insulator and helps keep these animals warm. Tiny shrimp-like animals called krill eat phytoplankton and algae under the surface of sea ice. They can also go without food for extended periods of time, allowing them to survive in the harsh winter months. Krill prosper in the millions and play an essential role in supporting the aquatic food chain by serving as food for whales, seals, penguins, and ice fish.

In the Middle East, camels have many adaptations that allow them to live successfully in harsh desert conditions. Long eyelashes protect their eyes from blowing sand, and thick eyebrows shield their eyes from the scorching desert sun. Slit-like nostrils can close to keep sand from entering the nose. Ears covered with hair inside and out keep sand and dust out. Large, flat feet help camels walk easily on the sand and not sink in. Fat storing "humps" allow them to go without food for several months, and their ability to go days without drinking water because of minimum water loss through urination and perspiration have supported camels in becoming an environmental success story. In Australian deserts, pooled water can be hard to come by, and so, thorny devils, which are diurnal reptiles, have developed spikes all over their bodies that are capable of absorbing water. Animals like octopuses, squids, chameleons, and cuttlefish rapidly change their colors to blend into the surroundings to avoid predators. Other animals like butterflies, deer, tigers, rabbits, leopards, etc., rely on their natural colors to blend into their surroundings, whether it is for hiding from prey or predators.

Time and time again, we have seen how changing environments have led to animals evolving physically and biotically to survive. Snakes, without much combative skills, evolved into venomous creatures to take down prey with little to no struggle. Some snakes, like the kingsnake, which are nonvenomous and eat other snakes, have developed immunity to snake venom. Giraffes have evolved from the now-extinct Samotheriums, which roamed the plains of Eurasia and Africa almost 7 million years ago. They had two ossicones on their heads and necks half the size of present day giraffes. It is said that the scarcity of food on the ground led to the elongation of their necks so that they could reach taller trees. Cave-dwelling animals such as some species of crabs and shrimps that lived underground where there was no light have lost the sense of sight over time. Researchers have found that the parts of their brains that control vision are becoming smaller while those associated with smell and touch are getting bigger. Another example of evolution because of climate change is that of owls. Tawny owls, which are prevalent in Finland, are found in two colors; pale grey and brown. Previously, more grey owls survived the winters as they were able to blend in with the snow, but over the last few decades, as winters have become milder, brown owls are now thriving as they are able to blend in with the branches of trees.

Then, there are extremophiles that flourish in extreme conditions like the deep sea. 200 meters below the ocean surface, with temperatures that fall below -5 degrees Celsius, extreme pressures, and no sunlight, deep-sea animals have evolved in unique ways to thrive in these extreme conditions. Due to the lack of sunlight, animals like the angler fish have developed the ability to create light through chemical reactions, also known as bioluminescence, in order to light their way and attract prey. Another fish that has evolved in the deep sea is the tripod fish, which is found 6000m below the ocean surface. They have long, bony fins which extend from the underside and allow the fish to “stand” on the ocean floor. They enable the fish to feel vibrations of prey on the ocean floor.

These numerous locales permit scientists and researchers to make comparable landscapes for life we might expect to encounter on other planets. We have almost unimaginable amounts of locomotion on the Blue Marble. Bipeds, quadrupeds, winged, slithering vertebrates, immobile giant clams consume proteins and sugars from the algae living in their tissues, flying mammals and reptiles. These constructs are configured for almost every fundamental structure: wings for flying in the air, legs or spines for walking across land, fins and articulated tails for swimming in water. Ironically, while man is the most intelligent being on Earth, I believe the most adapted is the duck. It can fly in the air, walk on land and swim in and under the water. While it is not amphibious, it certainly can mobilize itself through every elemental condition this planet presents.

Not just animals, human beings have also gone from being tree-dwelling anthropoids to the foremost intelligent beings on planet Earth. As the environment changed, savanna-like conditions encroached on the jungle environment, forcing ape-like ancestors to migrate into these new conditions to replace dissipating food sources. Going from eating nuts and berries, prey items became increasingly more dietary. This protein injection grew the cranial capacity and allowed for more cognitive development. As our cranial capacity grew, so did our intellect. What sets mankind apart from other living beings? One, the ability to conceptualize and 2, an opposed thumb. Living beings evolve according to environmental conditions. We grew opposable thumbs to grasp objects better.

The ability to figure things out and having the digits to explore and develop rudimentary technology were essential to our dominance of the planet. In only a few hundred years, we have gone from hunting animals for survival to landing on the moon. If we were capable of achieving all this in a few hundred years, there is no telling how advanced other beings, who have been around for millions of years, must be.

The number and nature of living beings found on the Earth alone are so diverse that one can only imagine what life on other planets, which are seemingly hundreds or thousands of years older than the Earth, could be like. There is a host of possibilities about life on another planet. A planet could be lifeless and also lack organic matter, or it could be lifeless but contain organic matter and fossils. At the same time, it could contain simple or complex life forms. The confirmation of any of these possibilities would be a massive step for science. What do the body types and configurations of extraterrestrial life forms look like?

Speculating about what aliens look like has kept children, filmmakers, and scientists and researchers occupied for decades. From personal accounts of people who claim to have seen aliens, and from the drawing and paintings of different cultures and religions who worship "sky Gods" that could possibly be extraterrestrials, the general description is that they are humanoids, four to five feet tall, with grey skin, a large head and large eyes. This implies that they probably come from a place with a low-light environment, which could also inherently imply that they live on a planet that closely resembles the Earth.

What are the general physical constraints on extraterrestrial life? While nothing is confirmed, considering the living conditions on Earth, scientists have drawn up some situations and physical attributes that extratropical might possess.

Chemically, life on Earth depends on carbon, nitrogen, and hydrogen, which use water as a solvent for biochemical reactions and a few others elements for survival. Phosphates, sulfur, and other elements are present in small amounts but are essential for survival. Would it be reasonable to assume that if life does exist on other planets, it depends on these same atoms and elements?

For there to be life on another planet, the structural molecules of living organisms should not be so stable that chemical reactions do not occur, nor should they be so unstable that the organism falls apart. Just as our molecules use water as an interaction medium, a medium for molecular interaction, preferably a liquid but possibly a dense gas, should be present. Life based on ammonia has been suggested as an alternative for life on other planets. For life to thrive in such conditions, the planet should therefore have some sort of liquid near the surface. If light from the sun is intense, then the surface should be shielded from this radiation. Lastly, environmental conditions should allow for autotrophy, which is the ability of an organism to produce its own food or some other means of producing necessary nutrients and compounds.

Thermodynamically, since photosynthetic organisms and the radiation they receive are not in thermodynamic equilibrium, extraterrestrial life forms might even survive through stellar radiation-based photosynthesis. Photosynthesis occurs when energy from a hotter object is transferred to a cooler object. On Earth, for example, a plant might have a temperature of 25 °C, but it is still synthesized by the sun, which has a temperature of about 5500 °C. If the source of radiation is cooler or has the same temperature as the synthesizing object, photosynthesis would not take place. Because of this reason, it is not improbable that a star could sustain photosynthetic life on other planets.

Where do aliens come from? While no

extraterrestrial life has ever been found on any planet as yet, some signs and possible conditions on other planets suggest that they might have once been inhabited by extraterrestrial life forms. NASA's spacecraft MESSENGER, which orbited Mercury between 2011 and 2015, found ice on Mercury. Scientists believe that parts of Mercury might have once been inhabited by microorganisms. It has long been speculated that alien life exists on Mars. The Curiosity rover provided evidence that strongly suggests the presence of a freshwater lake on Mars that could have provided a habitat for microbial life. Dried-up riverbeds, volcanoes, and polar ice caps suggest that Mars might have been warmer and wetter in the past and could have supported extraterrestrial life. Jupiter's moon Europa has long been the subject of speculation for the existence of life on it because of the high possibility of the presence of an ocean underneath its icy surface. If hydrothermal vents exist at the bottom of this ocean, they can warm the water and supply nutrients to microorganisms. Europa is considered to be the most likely planet that contains life, and a mission called "Europa Clipper" is planned for launch in 2024.

Other than these, many "super-Earths", which are planets twice the size of Earth and up to 10 times its mass, have been found outside our solar system. Since they are in the habitable zone and because of the presence of atmospheric water on them, super-Earths are considered to be the planets that are most likely to be harboring extraterrestrial life on them, as compared to smaller Earth-sized planets.

Circling back to the Fermi paradox, if extraterrestrial life forms exist and they are much more technologically advanced than us, why haven't they made contact yet? A new study has revealed what might be the answer to that question. While super-Earths are likely to be habitable because of their incredible mass, which gives them a stronger gravitational pull and allows them to better shield the surface from harmful cosmic rays, this strong gravitational pull might also make it more difficult for extraterrestrials to travel in any sort of spacecraft. "Civilizations from super-Earths are much less likely to explore the stars. Instead, they would be, to some extent, arrested on their home planet and, for example, make more use of lasers or radio telescopes for interstellar communication instead of sending probes or spaceships," says Michael Hippke, an independent researcher from Germany. One way for extraterrestrials to leave a planet more than ten times the size of the Earth would be to deploy nuclear force propulsion, in which a series of bombs is detonated in order to thrust a spacecraft through space.

A recent study about planet formation has revealed that more than 80% of the planets in our solar system already existed when the Earth was formed. If those many planets were already around in our solar system, there is no telling how many more existed in other solar systems. While there is no one factor that would help identify the presence of life, current or past, on any planet, scientists agree that a reliable biosignature for the detection of simple life on any planet would be the presence of oxygen. Any planet with more than 20% oxygen would be a good target for research. So, the first step to detecting the presence of life on other planets would be to search for oxygen and then follow that with the detection of other biosignatures. This can be achieved by deploying telescopes such as the European Extremely Large Telescope equipped with high dispersion spectrographs which are capable of detecting the oxygen lines in an exoplanet's spectrum.

In a nutshell, hundreds and thousands of planets orbiting the galaxies have been found, with more being discovered every other day. While life only positively exists on Earth, signs of past life and the possibility of life on some of these planets is high and, if research continues at the same pace, we might discover extraterrestrial life on one of them soon enough. Only time will tell when.

Chapter 9

The Day is Coming

Man is, if nothing else, the most curious being on the planet Earth. We wish to know all there is to know about our surroundings. While animals like dogs, dolphins, and monkeys are also curious, our superior intelligence is what sets us apart from all other living beings. Curiosity is so deeply rooted in our conscience that it helps us learn new things as we grow older and helps us survive as adults. “Babies learn more during their first three months of life than students do in four years,” says Alison Gopnik, an American Professor of Psychology at the University of California, Berkeley. The need to know how things work and why they are the way they are has been the driving force behind our development as individuals and as a species. “Why?” is a question unique only to humans. Why does the sun rise from the east? Why does it rain? Why do we look the way we do? Why are some people colorblind? Perhaps this curious trait is even *why* we are superior beings. Landing on the moon, discovering what lies in the depths of the ocean, determining where birds go in the winters, progress of technology, and the advent of Artificial Intelligence, all these would not have been possible if it weren't for our curious nature. We are in search of answers and simultaneously looking for ways to make our life easier.

While more research is still underway in many of these domains, such as the discovery of what lies in the deepest depths of the ocean and the exploration of space, we have come a long way from the primitive hunting ages, where our curious nature allowed us to design tools and devise plans to better hunt animals, where we came up with a way to communicate with each other through sounds and gestures. Even NASA'S largest and most capable rover, which was sent to Mars in 2012, was named curiosity. We are always looking for answers not only for things that are apparent but also for obscure things.

So, what is it, apart from curiosity, that drives us to explore broader horizons? What is it that keeps us going and has allowed us to discover amazing species and natural phenomena along the way? Whether it is exploring the Amazon rainforest, discovering what lies 2000 feet (the deepest human beings can go with the help of atmospheric suits) below the surface of the ocean, or what other planetary systems there might exist in the universe, one thing in common is our fascination with the unknown. Whether this is indeed fascination and not fear is up for debate.

When we are young, we are inquisitive and look for answers. If there are no answers to be found, we come up with an explanation ourselves. We do this because we do not like uncertainty. Uncertainty is scary. It makes us feel helpless and vulnerable. Fear of the unknown is nothing new; it has been around since the dawn of time. Why were we born? Why do we die? What is the reason for our existence? Before science, people used to rely on religion to explain the unnatural and the unexplained. The missing pieces of the mysterious puzzles of life and the universe were filled in by their beliefs which made them feel secure. Bertrand Russell said in a 1927 lecture, *"Religion is based primarily and mainly upon fear. It is partly the terror of the unknown and partly the wish to feel that you have a kind of elder brother who will stand by you in all your troubles and disputes. Fear is the basis of the whole thing – fear of the mysterious, fear of defeat, fear of death."*

Beliefs are important because they help us make sense of things and avoid uncertainty. Our beliefs shape us as individuals and influence our thoughts. Our need to steer clear of uncertainty is probably what gave rise to confirmation bias. It is very real and occurs when an individual believes in something even with little to no evidence to support it and also interprets new information in such a way that it supports their beliefs. The best examples of confirmation bias are probably how we see people today argue about whether climate change is real or not and whether the Earth is flat. The root of these arguments and beliefs is the same; our intrinsic fear of the unknown. Everyone has their own set of beliefs and assumptions that help them get through life, and so no one is willing to let go of them.

The nighttime sky has fascinated mankind since he first looked upward with clarity. When talking about the fear of the unknown, an imminent question that comes up while looking at the sky is, "Are we alone in the universe?" Considering the fact that the Earth is home to millions and billions of human beings and other species, people have always wondered if there could be more planets out there, just like Earth, home to millions of the same or different species. What conditions might they live in? Is contact with them possible? This curiosity led to the development and advancement of technology in the field of space exploration, and thus, since 1992, 4000 exoplanets and planets outside the solar system have been discovered, with the possibility of there being thousands more.

The possibility of there being life on other planets

has excited and intrigued not just scientists and researchers but also the common man for centuries. Where did the idea of other beings on other planets even come from? While many of the world's most peculiar and unique phenomena that once baffled everyone have been explained with the advancement in technology and research, there are still many things that are yet to be discovered and explained. While the first modern-day UFO sighting occurred in 1947, UFO sightings and other unexplained phenomena in the skies are not just limited to the last few years or decades; they have been happening for centuries. The first-ever recorded UFO sighting dates back to 1440 BC and was witnessed in ancient Egypt. The Tulli Papyrus is a transcription of an ancient papyrus from the reign of the Pharaoh Thutmose III (1504 – 1450 BC). According to the scribes of Thutmose III, "fiery disks" were seen in the skies. While the Tulli Papyrus has since been lost and historians and ufologists debate about its authenticity, UFO enthusiasts and book authors consider it as evidence of the first ancient UFO sighting.

EXOANTHROPOLOGY: A NEW SCIENCE FOR THE 21ST CENTURY'S FIRST CONTACT

While the ancient Egyptian sighting of the first-ever UFO may or may not be genuine, most of the medieval sightings, which can also be proved as accurate, come from Rome, as any event that occurred had to go through a series of rigorous procedures before it could be recorded officially. The famous Roman historian, Livy, recorded the sighting of a fleet of "phantom ships" floating in the sky in the winter of 218 BC. In 217 BC, Livy again reported the sighting of Parmas (round or oval shields made of iron or some other metal used by the Roman army), at the ancient city of Arpi. The next possible UFO sighting occurred a few years later, in 212 BC, when a giant boulder was seen flying about in the skies of Reate or Reiti. Just as in 218 BC, a fleet of ships was spotted in the sky in 173 BC in Lanuvium (modern-day Lanuvio). Weapons appeared to be flying in the sky in Compsa in 154 BC. The next reported sighting took place a few decades later, in 104 BC. The people of two Italian towns, Ameria/Amelia and Tuder/Todi, witnessed "weapons" in the sky rushing towards each other from east to west in what seemed to be a battle before disappearing into thin air. In 100 BC, a round shield, burning and emitting sparks, was observed in the sky, moving from west to east, somewhere in Rome. In 91 BC, the people of Spoleto witnessed a large, golden, ball-like object descend from the sky and take off again. According to the Greek historian Plutarch, thousands of soldiers of two Roman armies preparing for battle in 74 BC witnessed a giant, flaming object land between the two armies. It was shaped like a wine jar, and its color was like that of molten silver.

All the accounts of UFO sightings from the Roman Empire are considered to be genuine because they were witnessed by many people, sometimes by entire cities and armies. While it cannot be said for sure if what the ancient Romans saw were indeed alien spaceships or perhaps meteor showers or something of the sort, many of the descriptions, such as those of a flying object descending to the Earth and then rising up again, have no reasonable explanation other than that they were actually spaceships.

EXOANTHROPOLOGY: A NEW SCIENCE FOR THE 21ST CENTURY'S FIRST CONTACT

The list of centuries-old UFO sightings doesn't just end there, nor was it limited to the Roman Empire. Over the last few centuries, UFOs have been spotted in Korea, Japan, Mexico, and Europe. On April 14, 1561, what has been described as an aerial battle was witnessed by the residents of Nuremberg, Germany. Hundreds of cylinders, spheres, and other strange objects were seen floating in all directions in the skies. People also claimed to have seen a large, black, triangular object. This event has been termed "The Celestial Phenomenon over Nuremberg". A report printed by historian Samuel Coccis stated that strange phenomena occurred in the skies of Basel, Switzerland, two days in a row in the July of 1566 and again in August. In July, red and black balls seemed to come out of the sun and orbit it, and in August, black spheres covered the sun before breaking apart and vanishing. The next reported UFO sighting happened decades later, thousands of miles away in the Gangwon Province in Korea. According to records of the ruling Joseon Dynasty, on September 26, 1609, multiple witnesses in Goseong saw strange, flying objects similar to washbowls floating in the air in the early morning hours.

UFO sightings continued on into the 1800s, with multiple incidents happening in Japan, Mexico, and the United States. While some of these incidents were proved to have a logical explanation, others could not be explained, leaving everyone to assume that they really were UFO encounters. Maybe people weren't as curious about such phenomena in the earlier days because there was not much awareness about such incidents, but as technology has been advancing, we can easily access all sorts of information in seconds thanks to smartphones. Another reason for the growing interest in the existence of extraterrestrial beings is that UFO sightings have been increasing over the years, piquing everyone's curiosity.

Let's look at some huge incidents that have occurred in the last few decades that have no logical explanation and have strengthened our belief about the existence of aliens. During World War II, American pilots of the 415th Night Fighter Squadron reported seeing glowing, red lights in the air as they were flying.

The lights followed the fighter planes for approximately two minutes before disappearing into thin air. The lights were eventually named "foo fighters". Considered to be the first of the modern era of UFO sightings in the United States was the UFO sighting by private pilot Kenneth Arnold on June 24, 1947. Arnold claimed to have seen nine shiny UFOs at speeds of what he estimated to be 1200 miles an hour. He compared the shapes of these objects to that of a saucer or a disc, which led to the media coining the term "flying saucers". Arnold's story was validated by another witness who had viewed the same flying objects through a telescope at around the same time Arnold did. Soon after the Kenneth Arnold incident, numerous UFO sightings were reported in different areas around the United States. Whether they actually happened or were a result of people jumping the bandwagon is yet to be proved.

In 1948, US Air Force pilot George F. Gorman pursued a mysterious white orb at high altitudes in the skies of Fargo, North Dakota, for 27 minutes before losing it altogether. However, the first pictures to be taken of a “flying saucer” after the term was coined were taken by a couple on a farm in McMinnville, Oregon, in May of 1950. These photographs are considered to be the first genuine photographs of an unidentified flying object. In 1953, glowing red objects were witnessed by 45 residents of Rapid City, South Dakota. In 1961, the first widely publicized alien abduction incident occurred in a rural area of New Hampshire when a couple, Betty and Barney Hill, claimed to have been abducted and examined by extraterrestrials before being returned to Earth. In 1976, an object similar to a star was detected in radars over Tehran. Two F-4 Phantom II jet interceptors were sent to approach the object, but they reported the electronic equipment and commination being disabled as they approached the object. All connections were restored upon retreat. The most famous UFO sighting in Spain occurred in 1979 when a commercial flight was forced to make an emergency landing because of a bunch of red lights that appeared to be approaching the plane at high speeds. The phenomenon could not be explained. In 2006, 12 United Airlines employees reported seeing a saucer-shaped craft hovering over the Chicago O’Hare

Airport before departing suddenly.

The number of UFO sighting incidents around the world since the 1950s is huge. A UFO has been spotted somewhere in the world almost every single year since 1975, and the number of incidents has grown with every passing year. With lots of proof in the form of photographs and pictures, thanks to technology, it is hard to deny the existence of aliens because while some of these sightings have been invalidated, many still have no explanation other than that they are UFOs.

There are those who scoff at the incidents, stories, and sightings as hysteria or misidentification of other phenomena. The problem, however, is - they are too prevalent to overlook. The issue of extraterrestrial visitation has basically two camps: 1. Avid believers, 2. The dismissive. Both provide numerous data, facts, and postulates. I belong to the first camp, not because I want to believe (which I do) in extraterrestrial existence, but also because I think the empirical evidence is, if not overwhelming, at a little least whelming. How so? The sheer number of sightings, archeological sites where the mystery is unsolvable by our present scientists, e.g., Machu Picchu, videos from our own military aircraft gun cameras, and, yes, personal anecdotal accounts can only mean one of two things: (A) something beyond our science, knowledge, and understanding is going on, or (B) there is an exorbitant number of people with more psychological problems than can be comprehended. Either of these two is acceptable; therefore, the situation needs our ongoing interest and attention to determine if one or the other, or both, is occurring. If it is B instead of A, we're in bigger trouble than an unknown phenomenon. Since I believe it is A, the situation deserves more than a battle between speculation and naysaying. It is clear that the US military is taking it seriously. The question is: to what extent? At least from public reports, it is only a few personnel looking into the matter. So, what's next? Since the probability of extraterrestrial life forms on other planets is relatively high, and this cannot be denied, what is the next step as a society?

EXOANTHROPOLOGY: A NEW SCIENCE FOR THE 21ST CENTURY'S FIRST CONTACT

The first step is to prepare for formal contact with these beings. I do not believe any such preparations are being made by the government as we, as a society, still do not fully accept the presence and initiatives of other life forms. Since UFO sightings have been occurring for centuries now, it is not presumptuous to assume that these extraterrestrial beings wish to make contact with us; however, I have no knowledge of any government program preparing for such an eventuality. SETI (Search for Extraterrestrial Intelligence), CETI (Communication with Extraterrestrial Intelligence), and METI (Messaging Extraterrestrial Intelligence) are all relatively passive. Their goal is to detect any kind of signal from other planets; however, what would we do when we actually establish contact with such beings? While it was considered impossible once upon a time, scientists are now confident that contact with extraterrestrial beings is unavoidable and bound to happen eventually. Even the National Academy of Sciences has said that "Contact with other civilizations is no longer something beyond our dreams but a natural event in the history of mankind that will perhaps occur in the lifetime of many of us." There is no real groundwork for the eventual contact with extraterrestrial beings in place.

A group, or Scientific Task Force, is unquestionably the suitable undertaking. A combination of all categories of sciences and disciplines must be empaneled. To be sure, if extraterrestrials were to be contacted, we certainly would want to be the nation -out front- on reception. Demonstrated abilities of such visitations would not be in our best interests if hostile nations were the first to interact. I, therefore, propose this task force have physicists, medical doctors, perhaps animal biologists or even veterinarians, archeologists, exoanthropologists, military specialists in aviation and space experience, linguistics, communications technologists, psychologists, computer programmers (to put it all together in a usable form) and, dare I say it, reputable ufologists.

It is hard to deny that extraterrestrial beings are much more advanced than us. Some of the technology displayed by these entities is astonishing from currents physics as we know it. Maneuvers that, normally, would create inertial forces capable of splattering humans against the craft walls are somehow not an issue. Propulsion systems that move them around with impunity in our skies, seas, and near orbit space. Many, virtually all, can avoid interceptor crafts and radars as though they were toys. Yet, precious few encounters involve hostile actions. Indeed, none that appear the aggressor. Incidences of car engines shutting off and then restarting, jet planes losing navigational and other systems, but never propulsion, would indicate a high selectable technological imperative.

I would suggest that any course of analysis and study must include the actions to create a "Close Encounters of the Third Kind"-type of scenario. This must not be left to a 'someday' goal. Let's suppose we had a prepared site at the top of Devil's Tower or other remote location; from there, the broadcast of signals and language variations could be transmitted nightly in every technical form we could conceive. With all the advancements in technology, there are numerous mediums that can be utilized to establish contact. Audio, visual, telemetry, binary, laser, and other communicative forms would be sequentially sending 'We're Ready to Meet' messages. Some of these methods could be made mobile and sent to places where current 'flaps' occur. This is a major undertaking. Undoubtedly, significant government funding would be required, but that is what it's for! We have diplomatic outreach all over the world. Some off-world connections, especially if these connections could lead us into a new technological era, are more than worth the effort. We have nothing to lose and everything to gain from contact with extraterrestrial beings.

Lastly, one concern that arises from the attempts to establish contact with extraterrestrials is whether it is a good idea. While we cannot say what intentions they might have if we do establish contact, it is improbable that they would wish to cause us harm. This can be deduced from the fact that all the "abductions" that have occurred have always resulted in the return of the abducted subjects. The term “abduction” has sociological importance. We abduct hundreds and thousands of animals each year for experimentation, and that is considered normal; however, if one of our own is abducted, we think of it as a travesty. We have all these laws and rules in place to benefit our causes. Perhaps these extraterrestrial beings also have their own set of rules in place, which determine what they can and cannot do with us if they are to come in contact with us.

I would end this chapter by saying that there are endless possibilities to any new exploration or discovery, contact with extraterrestrials being some of them. While we cannot know for sure what the outcome of such contact might be, we must be prepared for the eventuality of it happening and so must be proactive and prepare for it beforehand.

Chapter 10

Epilogue: Where Do We Go from Here?

"In very different ways, the possibility that the universe is teeming with life, and the opposite possibility that we are totally alone, are equally exciting. Either way, the urge to know more about the universe seems to me irresistible, and I cannot imagine that anybody of truly poetic sensibility could disagree."

- Richard Dawkins

It is incredible how people can close their minds off to the size of the Universe. With billions of stars, millions of galaxies, and hundreds of thousands of planets, with more being discovered routinely, how can it be that human beings are the only thinking beings in creation?

As mentioned before, there are two camps when it comes to the existence of life on other planets; the believers and the naysayers. The most glaring contradiction about life in the Universe is: Where do the naysayers *think* we are? An insignificant yellow star on the outer rim of the Milky Way Galaxy in a Universe of trillions and trillions of stars, I believe, makes a case for advanced civilizations from these simple, logical perspectives: 1. Earth not only has life but millions of species and forms. 2. Trillions upon trillions of stars operate the same way everywhere in the Universe. 3. The celestial mechanics that put life here are likely proceeding everywhere. 4. Earth formed 4 billion years ago, but the Universe has been evolving for 14 billion years! Nearly 10 billion years of evolution and 'progress' have occurred before us!

Below I have inserted some information from the Russian scientist Kardashev, discussing the three levels of technological civilizations, which have now been expanded to seven levels. This is known as the Kardashev Scale and is a method of measuring a civilization's technological advancement level based on the amount of energy that it is able to use.

Type 0 (Sub global Civilization): A civilization that can utilize the energy of its home planet, but not to its full potential. We fall under this category. Currently, we are at 0.73 on the Kardashev scale. Scientists have proposed that we will reach Type I in about a hundred or two more years, depending on the speed of advancement of our technology.

Type I (Planetary Civilization): Planets that fall under this category can use and store all the energy that falls on the planet from its parent star. This would mean utilizing the solar energy that comes from the star and also the other kinds of energy (thermal, hydro, etc.) produced on the planet itself. To reach Type I would mean to be able to control all aspects of the planet, including influencing the weather, controlling earthquakes and volcano eruptions, influencing flora and fauna, etc.

Type II (Stellar Civilization): A civilization that has control over the solar system and is capable of using the entire radiation output of its star. This can be achieved by a Dyson Sphere, a structure that can enclose a star and capture all its energy.

Type III (Galactic Civilization): A galactic civilization can harness, store, and control the energy output of all the stars within a galaxy (almost 10 billion times the energy output of a Type II civilization). It would probably take us 100,000 years or more to reach this level.

Type IV (Universal Civilization): This civilization would be capable of harnessing the energy of the entire universe. Such societies could possibly become immortal and achieve energy-matter transformation, teleportation, and time travel.

Type V (Multiverse Civilization): This kind of civilization would be able to explore the multiverse and harness the energy of not one but multiple universes.

Type VI: A type VI civilization would be beyond the

multiverse, in the Megaverse, existing outside of time and space and capable of creating and destroying universes and multiverses.

While the sheer scale of achievement is gigantic compared to what humans can now do, I would point out the Universe is just as large. Ever since we appeared on this planet, we have been coming up with solutions to problems and technologies to make life easier. We have gone from living in caves to going to space. So, although it seems impossible right now, it is not to say that we cannot reach these levels someday, albeit hundreds or thousands of years later.

A micro-dimensional extension to the Kardashev Scale was proposed by astrophysicist John D. Barrow, and I think his -minus variations are very realistic to clarify these scales most accurately. Instead of basing the levels of technological civilizations on the energy utilized, he instead based them on the smallest structures that the civilizations can manipulate. Listed below are the levels on the Barrow scale:

Type I-minus: This society can control and

manipulate large-scale objects that they can see and touch, such as building structures, mining, etc. Human history, until the mid-nineteenth century, falls in this category.

Type II-minus: This society is capable of manipulating genes and macromolecules. Examples include genetic engineering and organ transplantation.

Type III-minus: This type of civilization can manipulate molecules and molecular bonds in order to create new materials or metamaterials.

Type IV-minus: This level involves the manipulation of individual atoms and the creation of nanotechnologies, both of which can result in the ability to create artificial life.

Type V-minus: This level is capable of manipulating individual atomic nuclei, allowing for the transmutation of elements.

Type VI-minus: This level can manipulate the most fundamental particles of matter (quarks and leptons).

Type Omega-minus: This level can manipulate the basic structure of time and space.

The human civilization is somewhere between Type III-minus and Type IV-minus according to this classification.

Manipulating the very structure of the Universe is no small chore, but UFO observations and reports make it clear that someone has developed technology to such a level. This is more proof that Exoanthropology and its' established and emerging sciences must combine to prepare for interacting with beings of this level of advancement. We do not want to be a Neanderthal listening to a millennial describe the working of his cell phone!

There are many mysterious phenomena and mysterious sightings throughout history that would suggest the existence of something extraterrestrial in nature out there. Even in the world around us, there are so many mysteries that cannot be explained. Take the Puma Punku, for example. Constructed between 536 – 600 AD (although scientists argue that it was built much earlier than that), the Pumapunku or Puma Punku is a large temple complex near Tiwanaku, Bolivia, part of a larger archaeological site known as Tiauanacu. What was once a terraced platform mound with a sunken court built with megalithic blocks, each weighing several tens of tons, now lies in ruins. However, the mystery of the Puma Punku has continued to baffle scientists and archaeologists for centuries. Spanish travelers in the 16th and 17th centuries described it as “a wondrous, though unfinished, building with gateways and windows carved from singe blocks.” To explain what has intrigued and amazed researchers about the construction of the Puma Punku, here is an excerpt from an article about the engineering involved in the temple’s construction: “In assembling the walls of Puma Punku, each stone was finely cut to interlock with the surrounding stones, and the blocks fit together like a puzzle, forming load-bearing joints without the use of mortar. One common engineering technique involves cutting the top of the lower stone at a certain angle and placing another stone on top of it, which was cut at the same angle. The precision with which these angles have been utilized to create flush joints is indicative of highly sophisticated knowledge of stone-cutting and a thorough understanding of descriptive geometry. Many of the joints are so precise that not even a razor blade will fit between the stones. Much of the masonry is characterized by accurately cut rectilinear blocks of such uniformity that they could be interchanged for one another while maintaining a level surface and even joints."

The precision displayed in the cutting of the blocks also suggests the possibility of prefabrication and mass production, which, coming from a civilization that had no writing system and did not even know about the wheel, are technologies far too advanced to have been achieved by them. Numerous modern-day engineers are baffled as they cannot duplicate the precision, allegedly done by early man with hammers and chisels, even with modern state-of-the-art machinery.

Apart from the skillful carving of the blocks, the size itself of these blocks was enormous. The largest block found at this site is 25.6 feet long, 17 feet wide, 2.5 feet thick, and weighs around 131 metric tonnes. The second-largest block weighs 85 metric tonnes. This also raises the question of how these massive blocks were transported to the temple location as it is high up in the Andes Mountains. Moreover, the temple is located in an area above the tree line, meaning there were no trees that could have been cut down to be made into rollers or other instruments of transportation. It has still not been determined what tools were used for the precise cuts and shapes that were carved into the stones. Examples of the shapes and marking found on the stones include perfectly straight lines, perfect holes, H-shaped blocks, and cubes. To add to the mystique of these puzzling structures, they are also magnetic!

Mainstream archaeologists accept this amazing site's construction, without any consideration of the building techniques, as being made by ancient humans with hand tools! Ground moving that only bulldozers and graders could accomplish today. Edges are, after apparently thousands of years, still razor-sharp.

Diamond saw equivalent shaping with laser-like alignment, just not a reasonable "explanation" of mere manual labor! I believe this is the SINGLE MOST evidentiary material demonstration of non-terrain engineering expressed on Earth. Conquistador's inquiry of indigenous Incans of the building techniques was told, "these were made by the Gods, not us".

There are many sites around the world that people believe were built by aliens, including the ancient Egyptian pyramids. The same questions arise about the construction of the pyramids. How were they made? How were the huge blocks transported? But perhaps more surprising is the fact that the three pyramids surrounding the Great Sphinx; the Pyramid of Menkaure, the Pyramid of Khafre, and the Great Pyramid of Giza, are perfectly aligned with stars in Orion's belt, and this perfect alignment has long been considered a "gateway to the stars". Even though we know the Egyptians were a much more advanced civilization than most in architecture, art, construction, agriculture, medicine, and technology, the construction of the pyramids still seems much too advanced even for them. While the pyramids may or may not have been built by humans is debatable, there is no real explanation for how the construction of Puma Punku was achieved by a primitive civilization.

Humanity has, over its existence, grown long legs in societal and cultural norms. Our own history makes significant points that still haunt and curse us today. Curiously, this same species still has elements that have ruled our lives for eons. Before the establishment of governments, religion was the main societal organizing force. The problem with religion, however, is that it is based on faith and not reality. When the Renaissance arrived, there was the awaking to the natural world. There was an emphasis on humanism and individuality. Self-appointed experts gave us incorrect and misleading information until it was abundantly clear that a natural and, more importantly, accurate view of existence was revealed. This even led to the inhibition of scientific findings as late as the 15th century. Even today, there are religious fanatics who actually kill other people because their belief system is rejected. More importantly, those who have made the most significant strides toward the future of humanity are periodic geniuses. The Einsteins, Newtons, Salks, and Teslas of the world give us advancement! Over time, it has been revealed that when humanity questions the status quo, we discover so much more.

I recently read an article that claimed the theory about aliens constructing advanced buildings in ancient times is racist because it questions the technological achievements made during a savage period of earlier human civilizations because of their skin color. Insane. Facts are not 'racist'. If that writer can explain why engineers with the latest technology cannot DUPLICATE certain monolithic constructions and how the Puma Punku was built, I would be more than happy to learn. This is the problem with establishment 'knowledge'. They say no, no, no, yet they cannot explain how razor-sharp edges on stones during a stone hammer and, at best, copper chisels period, could execute flawless constructs that diamond saws and lasers cannot do today!

Clearly, some kind of external influence and knowledge beyond human comprehension was involved.

During the Renaissance, which took place from the 14th to the 17th century, astronomy and science got their collective divisions. Man is the most curious being on the planet. It is in our nature to seek more knowledge and know all that there is to know. I believe the chemistry, that is us, goes beyond mere existing and actually has a yearning for our stellar heritage intrinsically built-in. For a long time, science has proven that we are made of star-stuff. The atoms of our bodies are forged in stellar foundries and ultimately came to establish our human form. Yet, we struggle against our own composition. Emotions can distract our thoughts from more progressive ideas. Politics, war, oppression, and other human failures keep us distracted from advancement that may even come further and faster than some other beings who have 'evolved' past those shortcomings. That reptilian ancestry, buried deep at the base of our skulls, confounds many. This may be an indication that a technocracy form of social order will, one day, prevail.

We are faced with many daily challenges; however, our future is certain. One day, the star that is our Sun will use up its nuclear fuel and begin to swell into a Red Giant. This inevitable solar event, destined to stars the size of ours, will consume the inner planets of which we are the third. Long before that, we will hopefully have made an exodus to a new planet(s) where we can continue as a species. Science and technology will get us there. Politics, prejudices, warfare, religious objection, and other narrow-minded thinking will not.

Another conspiracy theory out there is that aliens do, indeed, exist, but their existence has been hidden from the general public by the governments for whatever reasons. Ever since 1947, cover-ups have been the staple of UFO lore. Thoughtful analysis, back then, was probably correct to have "plausible denial" of any potential extraterrestrial activity. After the fiasco of the "War of the Worlds" radio broadcast by Orson Welles, less than a decade earlier, caused mass hysteria and panic, governmental officials certainly took note.

However, sometimes, the government has a problem with overreaction and prolonged overreach. Now, in the second decade of the 21st century, some rethinking must be done. Of course, no thinking citizen would want national security to suffer over curiosity and wonder. Yet, this is easily solvable. First, some hope has been demonstrated by the recent release of the US Navy aviation gun camera/radar footage of some highly unusual craft operating in restricted airspace. The release was well greeted by UFO enthusiasts who did not complain that the footage was 6 to 8 years old! This is demonstrable of a way forward. It was also noted that the AATIP Program was a funded UFO investigative program for crafts the government swears do not exist. Actually, this is the kind of scam that angers UFO enthusiasts who cry over cover-up and ordinary citizens who voice a lack of faith in government transparency and truthfulness. What should happen is: A government spokes agency be designated and relate current views and speculation on what is occurring and be truthful about what is known. This does not, nor should it, require national security level data and information. For example, if there *is* material and, yes, even expired occupants on-hand, so state that our foreign adversaries cannot be allowed to know all we know and such material/related information is not going to be released. Clear, concise and explained with proper reasoning.

Now that we know there is ample evidence to suggest that extraterrestrial life, far more advanced than us, has been around for hundreds of years, what is the way forward? The next step is to establish contact with these other worldly beings ultimately. For that, exoanthropology needs to be the central area of focus, primarily because scientific analysis alone does not give us any information about these beings' culture or fixed medium of communication and to approach them, I believe the contact must be structured from numerous points, simultaneously. Firstly, *how* do they communicate among themselves? There is much anecdotal evidence that they use multifaceted methods during human interaction. They have responded to flashing lights with flashing lights. Some seem to have telepathic abilities. Examining various bio symmetry, we see some have ears, some have little to none. Is it due to lack of need or use? Some seem to know local tongues in ancient times as no mention in recorded history EVER mentioned a language issue. Like how the Annunaki were able to communicate with ancient Sumerians. My suggested multi-discipline working groups may wish to develop various languages from around the globe and computerize a "lingual bank" that can immediately respond to a recognized dialect. Some study and include HOW was Sumerian actually spoken.

This is not an easy feat to achieve, of course. Creating a new branch of science will always be challenging. Fortunately, we have many scientific disciplines that have evolved over time and, hopefully, this will bode well for more open-mindedness toward this goal. As we have learned as humans, the first answer isn't always the correct one, such as the Earth is flat; the sun revolves around the Earth, and man cannot exceed the speed of sound without dying. This same open-mindedness is the hallmark of genuine research and advancement! Contact is bound to happen one day. Scholarly pursuit now will prevent mistakes and allow us to have countermeasures so that a greeting is not an insult. Mystery and curiosity go hand in hand. It is an essential element of our human makeup. Why do we want to go into space? Simply because it is there, and we want to know what it encompasses.

I think the first pursuit in the field of Exoanthropology should be topics and experiences molded into curricula. Filling the curricula can come from many sources. We must expect that some data will be pure speculation. Still, some very credible data has also been collected over the years and translating that into subject matter will come from amateur or novice subject matter experts. Many UFO books come from readily identifiable personalities such as ones found on the Ancient Aliens series, and let's not forget people who have experienced some form of contact or abduction with credible scenarios and witnesses. They would be called upon by Curricula Formulators and develop an academic strategy from this gathering of information from individuals. There will undoubtedly be a conflict between different camps, and strong conference leaders will be needed to extract usable approaches from less than credible details and concentrate on a "Prove or Disprove" methodology.

Again, starting with a "blank piece of paper" is always best. Avoid prejudices, bias and other "established" ideas that have proven to be, ultimately, "wrong" yet supported for decades or longer! I believe the motivating force will be those who wish to understand that we must do something to get where we need to be in preparation for contact as we don't want to be just standing there, appearing clueless and not considered worthy of introduction!

It is clear that establishment ideas die hard. Hopefully, a series of Conferences and Seminars would generate a "common goal" attitude with younger yet brilliant scholars from all the scientific fields that will be needed to make this a rational academic pursuit. The Government could help immensely by noting that they are NOT discounting such knowledge. It must be researched, and a consensus of outcome will be valuable to us not only as a nation but for the entire human race. Some members of the intelligence community willing to support this line of study would be encouraging. In short, we must approach this topic as we would any other requiring investigation; through analysis and, ultimately, protocols. This does appear not to be the 'undignified discussion' it was just a few years ago. The release of navy gun camera video of craft displaying unknown technologies was admitted by the Government. Something is out there! The question for contrary Academia is, "Do we bury our heads in the sand OR take a fresh, new approach and give it the analysis and attention it deserves in the ongoing study?"

Convincing humans of anything, initially, is difficult. Here, we have an opportunity to formalize a new in-depth and, most importantly, serious ongoing study of an esoteric subject that has proven that it *must* be studied and yet has not garnered the attention it deserves from the educational community. This MUST change. Hopefully, this book is one of the "first shots" in the war against the unknown.

When this idea finally comes to fruition, there is no telling what we will discover and what we will accomplish as the human race. We can only wait as we work towards advancing our technologies enough to establish contact with extraterrestrial beings, and the story will write itself from there onwards.

Made in the USA
Monee, IL
04 November 2026

67380864R00108